Blackhatonomics

Blackhatonomics

An Inside Look at the Economics of Cybercrime

Will Gragido

Daniel Molina

John Pirc

Nick Selby

Andrew Hay, Technical Editor

ELSEVIER

AMSTERDAM • BOSTON • HEIDELBERG • LONDON
NEW YORK • OXFORD • PARIS • SAN DIEGO
SAN FRANCISCO • SINGAPORE • SYDNEY • TOKYO

SYNGRESS.

Syngress is an Imprint of Elsevier

Acquiring Editor:	Steve Elliot
Development Editor:	Heather Scherer
Project Manager:	Malathi Samayan
Designer:	Mark Rogers

Syngress is an imprint of Elsevier
225 Wyman Street, Waltham, MA 02451, USA

Library of Congress Cataloging-in-Publication Data
Gragido, Will.
Blackhatonomics : an inside look at the economics of cybercrime / Will Gragido [and four others].
     pages cm
Includes bibliographical references and index.
ISBN 978-1-59749-740-4
1. Computer crimes–Economic aspects. 2. Cyberterrorism. I. Title.
HV6773.G7237 2012
364.16'8–dc23

                                        2012043186

British Library Cataloguing-in-Publication Data
A catalogue record for this book is available from the British Library

ISBN: 978-1-59749-740-4

Printed in the United States of America

Transferred to Digital Printing in 2014

Working together
to grow libraries in
developing countries

www.elsevier.com • www.bookaid.org

For information on all Syngress publications, visit our werbsite at *www.syngress.com*

Acknowledgements

I'd like to thank my wife, Tracy for her unfailing support in all that I do, my children and my co-authors Dan, Nick and John. I'd also like to thank Red Bull.

—Will Gragido

First and foremost, I want to thank my family for enduring the writing process, and specifically my wife, Lisette, for it was her support during this past two years as the project developed and evolved that kept me motivated and progressing. To my daughter, Danielle, you can always make me laugh, and keep me humble. To my co-authors, you have shared and shown me a lot, and I appreciate your letting me share this experience with you. To my friends and colleagues in the Information Security field, starting with Chris Klaus and Tom Noonan, who gave many of us a chance to learn the basics back in the ISS days, to Gary Mullen, and Ray Menard, who have been great colleagues and even better friends. To my Kaspersky Lab team, those in ELAM, LTAM, those in the GReAT team, and those in Emerging Markets, thank you for growing and learning with me. Lastly, to Eugene Kaspersky, who showed me that your passion can drive your life's goals far beyond what one single human can dream.

—Daniel Molina

To my boss Greg Adams, thank you for believing and encouraging me. Your leadership and examples you demonstrate on a daily basis is one that I will never forget. To my best friends Eric York, Heath Peyton and Will Gragido… so thankful to have you guys in my life and thank you for the constant encouragement during a rough time! To my mom Judy Pirc and my sister Jamie Line, thank you for your love and encouragement. To my team at TippingPoint (Orion Suydam, Chris Thomas, Dave de Valk (Virtual Dave), James Collinge, Patrick Hill, Pat Geistman and Rohan Kotain), thank you for the honor to manage you guys…you taught me a lot in very short period of time. Chris Haskins, Bobby and Veronica Gideon, Charles and Crystal Carlson, Scott Lupfer, Brian Reed, Dinesh Vakharia, Chuck Maples, Jay Rollete, Scott Rivers, Mike Polston, Rhonda Pouraty, Jon Dykes, Chris Radosh, Kathy Skinner, John Trollinger & Chad, Ryan Strecker, Sanjay Raja, Reese Ann Sims, Dan Holden and Jennifer Lake…

thank you as each one of you invested time in me. DJ, Nick and Will, it was a pleasure to work with you all on this project. I have tremendous respect for each of you and thank you for an incredible knowledge transfer. To my children, Kelsey, Aubrey and Jack, I love each one of you more than anything in world and thank you for supporting your daddy! Lastly, I want to thank my Lord and Savior Jesus Christ as none of this would have been possible without HIM (Proverbs 3 5:6). WYMM?

—John Pirc

Thank you to my co-authors for getting me involved with this project, and your inspiration and support through the process. Thank you to Bill and Natalie, and officers, defense and prosecution attorneys in the US and Europe for your input to and peer-review of the chapter on global law enforcement. To the small international cadre of law enforcement officers who combat uneducated policymakers, unmotivated and uneducated officials and slog across politics-mined terrain as they carry out their mission of investigating cyber crimes and prosecuting cyber criminals: to quote a friend in public service, "Thank you so much for your thankless work."

—Nick Selby

Dedication

This book is dedicated to all those seeking to comprehend the nuances associated with and pertaining to, illicit sub-economic markets; particularly those associated with cyberspace.

—Will Gragido

This book is dedicated to the security professionals that are surviving the ever-changing realities of cyber-crime, cyber-terrorism, and cyber-warfare in today's world.

—Daniel Molina

I want to dedicate my portion of the book to Nicole Langenbach. Nicole, I'm very thankful that the Lord placed you in my life at the right time. I'm also looking forward to spending the rest of my life with you. I also want to thank Ted Ross and Jennifer Parker for re-introducing me to you.

—John Pirc

For Corinna and Spijk.

—Nick Selby

Contents

About the Authors

Will Gragido possesses over 18 years of information security experience. A former United States Marine, Mr. Gragido began his career in the data communications information security and intelligence communities. After USMC, Mr. Gragido worked within several information security consultancy roles performing and leading red teaming, penetration testing, incident response, security assessments, ethical hacking, malware analysis and risk management program development. Mr. Gragido has worked with a variety of industry leading research organizations including International Network Services, Internet Security Systems/IBM Internet Security Systems X-Force, Damballa, Cassandra Security, HP DVLabs, and now RSA NetWitness, where he leads the RSA FirstWatch Advanced Threat Intelligence team.

Will has deep expertise and knowledge in operations, analysis, management, professional services & consultancy, pre-sale/architecture and strong desire to see the industry mature and enterprises & individuals become more secure. Will is a long-standing member of the ISC2, ISACA, and ISSA. Mr.Gragido holds the CISSP and CISA certifications, as well as accreditations in the National Security Agency's Information Security Assessment Methodology (IAM) and Information Security Evaluation Methodology (IEM). Additionally, Mr.Gragido is a Faculty Member of the IANS Institute where he specializes in advanced threat, botnet, and malware analysis. Mr.Gragido is a graduate of DePaul University and is currently preparing for graduate school. An internationally sought after speaker, Will is the co-author of *Cybercrime and Espionage: An Analysis of Subversive Multi-Vector Threats*.

Daniel J. Molina (CISSP) is Senior Regional Director for ELAM (Emerging Latin American Markets) in Kaspersky Lab. In this position, he serves as a managing director for the region, inclusive of sales, marketing, channel development, engineering and support.

Mr. Molina is considered a thought leader in the area of information security, and has been called to speak on issues such as the state of the security industry,

"Security Best Practices", "The Business Aspects to Information Security", "Operational Efficiency in IT Security" and "The Myth of ROI in Security", and "Capabilities Maturity Models in Security" at various industry forums worldwide. His view on security maturity has made him a sought-after resource to help explain and justify, in business terms, what users, businesses, and government entities require.

Daniel was previously Channel Sales Director for Latin America and the Caribbean at Kaspersky Lab, and first joined as part of the Field Marketing team for part of the United States.

In his previous role as Director of Advanced Solutions, and as a Security Evangelist for McAfee, Mr. Molina provided a voice to the McAfee Risk Management Process, and assists in complex and strategic opportunities for McAfee customers. He has also created curriculums, and provided training to multiple partners on Security Intelligence, and Anomaly Detection and Behavioral Forecasting models for security.

Daniel has extensive experience in enterprise security architecture design, internetworking, LAN/WAN implementation and project and team management. In addition to his role at McAfee, Daniel spent several years as a Principal Systems Architect for Q1 Labs, Solution Architect for Internet Security Systems and as an Enterprise Consultant with Entex Information Services overseeing infrastructure and Y2K project implementations for companies such as GTE, Nextel, and The Coca-Cola Company.

Daniel's background includes several years as a systems specialist and administrator with enterprise and carrier environments. Along with numerous industry and technology-specific certifications, Daniel holds the following designations: CISSP, CBS, CCSA, CCSE, MCSE+I, and others.

Daniel studied Political Science and Psychology at the University of Southern California and Economics at the University of Texas, Arlington.

John Pirc recently co-authored *Cyber Crime and Espionage*, published in February 2011. He has more than 15 years of experience in Security R&D, worldwide security product management, marketing, testing, forensics, consulting, and critical infrastructure architecting and deployment. Additionally, John is an advisor to HP's CISO on Cyber Security and lectured at the US Naval Post Graduate School.

John extensive expertise in the Security field stems from past work experience with the Central Intelligence Agency in Cyber Security, as Chief Technology Officer at CSG LTD, Product Manager at Cisco, Product Line Executive for all security products at IBM Internet Security Systems, Director at McAfee's Network Defense Business Unit and currently the Director of

Product Management at HP Enterprise Security Products leading the strategy for the organization's next generation security platforms.

In addition to a BBA from the University of Texas, John also holds the NSA-IAM and CEH certifications. He has been named security thought leader from SANS Institute and speaks at top tier security conferences worldwide and is most recently published in Forbes on Social Media security.

Nick Selby has been an information security analyst and consultant for more than a decade, and has worked in physical security and intelligence consulting in various roles since 1993. In 2005 he established the information security practice at industry analyst firm The 451 Group, where he conducted in-depth technical briefings and consulted more than 1000 technology vendors. Nick has consulted hundreds of venture-backed startups on understanding their competitive landscape, on product development and feature enhancements, user interface and security. He has consulted US and European governments, more than 80 investment banks, more than 20 venture capital firms; on the investment side, to better understand the technology and landscape of the companies into which they invested, and on the operations side on securing their intellectual property and processes. In 2007 he was appointed VP of Research Operations at 451, where he managed more than 35 technology analysts, developing analysis products and technologies to leverage their insights. Since 2006 Selby has served on the faculty of IANS Research. His work consulting F500 companies on data theft and industrial espionage has placed him at the leading edge of firms helping those under attack by adaptive, persistent adversaries, and he is experienced at managing attacks and architecting recovery networks.

Since 2008 he has focused on law enforcement intelligence, and he works part-time as a sworn police officer in the Dallas-Fort Worth Metroplex, investigating cyber crime. He teaches continuing legal education on cyber crime for prosecutors in one of the country's largest jurisdictions, and writes the TechTalk column for Law Officer Magazine. He is the CEO of StreetCred Software, which produces software that helps law enforcement serve fugitive arrest warrants through predictive intelligence.

About the Technical Editor

Andrew Hay is the Chief Evangelist at CloudPassage, Inc. where he serves as the public face of the company and its cloud server security product portfolio. Prior to joining CloudPassage, Andrew served as a Senior Security Analyst for 451 Research's Enterprise Security Practice (ESP) providing technology vendors, private equity firms, venture capitalists and end users with strategic advisory services—including competitive research, new product and go-to-market positioning, investment due diligence and tactical partnership, and M&A strategy. Through his work at 451 Research, Andrew was instrumental in securing tens of millions of dollars in equity investment for numerous security product vendors. He is a veteran strategist with more than a decade of experience related to endpoint, network and security management across various product sectors, including security information and event management (SIEM); log management; deep packet inspection (DPI); security analytics; vulnerability management; penetration testing; intrusion detection and prevention (IDS/IPS); firewall; threat intelligence; application whitelisting; network and host forensics; incident response; and governance, risk and compliance (GRC).

Before joining The 451 Group, Andrew worked in the Information Security Office (ISO) of the University of Lethbridge, in Alberta, Canada and, prior to that, at a privately held bank in Hamilton, Bermuda; in each position, he was responsible for strategically designing, driving and executing the goals and objectives of the organization's information security programs. Prior to that, Andrew served in various roles at Q1 Labs, including Engineering Manager, Product Manager and finally as the Program Manager responsible for the entire portfolio of third-party technology partner relationships.

Andrew was honored with the title of Security Thought Leader in May 2008 by the SANS Institute; named an IT Knowledge Exchange blogger of the week in June 2009; listed as one of the 10 'Infosec Folk to Follow on the Twitters' by Matthew Grant in November 2010; listed as one of the Most Powerful Voices in Security by SYS-CON Media's Jim Kaskade in September 2011; and named one of Tripwire Inc.'s Top 25 Influencers in Security in December 2011. Andrew was also presented with the Lethal Forensicator Coin by The SANS Institute at the SANS 2010 What Works in Forensics and Incident Response Summit—awarded to those who demonstrate exceptional talent, contributions, or helps to lead in the digital forensics profession and community. He is a sought-after speaker, and has presented at numerous international security conferences such as the SOURCE Conference, Infosecurity Europe, SANS What Works in Forensics and Incident Response Summit, SANS Network Security, Security BSides, RSA Security Conference, Americas Growth Capital and the joint iTrust and PST Conferences on Privacy, Trust Management and Security.

Andrew is frequently approached to provide expert commentary on security-industry developments, and has been interviewed by members of the press for such publications as The Sacramento Bee, eWeek, TechTarget, Wired Magazine, Network World and CSO Magazine, in addition to podcasts such as the Data Security Podcast, Forensic4Cast, SecuraBit, PaulDotCom, Security.Exe, Beyond The Perimeter, The Risk Hose, Security Roundtable and Tenable Network Security. He was formally the founder and cohost of the LogChat podcast with Dr. Anton Chuvakin. Andrew also has written articles for several trade publications such as Information Week Magazine, DarkReading and Network Computing on various security-related topics.

Foreword

Would you be surprised if I say that the essence of cyber-crime comes down to traditional theft of property—be that of funds in accounts, of company data (and the financial harm that can cause), or of personal/confidential information (and the damage that can cause by it winding up in the wrong hands)? As time goes by, hackers, virus writers and other cyber-criminals are gaining more and more experience and thereby increasing their sophistication, highlighted by their employing dirty tricks in their illegal endeavors (for example, using legal digital certificates in malware). Obviously, corresponding defenses need to increase in sophistication too. To do that we need to get into the minds of cyber-criminals to understand their motives and the economics at play in cyber-crime. This book helps to do that. The real motivation of a cyber-criminal is explained in simple economic terms. No acronyms relating to worms and Trojans, just the underlying facts regarding the global business of cybercrime, the reasons involved, and the players taking part (including real world cyber-crime case studies and interviews with high-profile cyber-criminals)—through the experienced eyes of the expert authors.

The threats that stem from cyber-warfare are several: (i) if cyber-warfare weapons are uncovered—as they occasionally are (Stuxnet, Duqu, Flame, Gauss)—they are easy to copy and modify, including ultimately by cyber-terrorists, who, unlike nation states, will never be held to account by any worldwide non-proliferation treaties that may come about in the future; (ii) cyber-warfare weapons can have unintentional side effects affecting the same hardware or software right around the world—hardware or software that might control critical life-supporting infrastructure (think reservoirs, the power grid, the food chain); (iii) working out who is behind a cyber-war attack is practically impossible, meaning cyber-warfare weapons will continue to be used a lot—since those behind them can get away with it with no consequences, for example retaliation; and (iv) cyber-war weapons are, by definition, nation state-backed, meaning the research that goes into them is very well financed, meaning the resultant sophistication of cyber-weapons is state-of-the-art, and as a consequence may be applied undetected for years. Again, this book comes up with answers, backed up by real life examples, to help us understand what needs doing to be better able to deal with the serious global cyber-war threats posed by the digital free-fire zone that is fast becoming today's reality.

Should we be concerned by growing cyber-crime and cyber-warfare? I certainly think so, but then—you could easily say—I would: I sell security products. If you don't believe me, read this book, then see. Regarding the latter—cyber-warfare, opinion is still rather mixed about the real dangers posed. This is because there is little empirical evidence to be found, while academic and policy papers largely rely on anecdotal case studies. Once again, this is where this book comes in.

The authors give their expert insight on the road ahead as it relates to the many facets of cyber-security, including but not limited to industrial espionage, the economics of security and the geo-political landscape. They also comment on issues within the security community today that act as impediments to any organization worldwide achieving the security needed as it takes on the challenges of the road ahead.

The authors have between them several decades of experience in security while at the same time come from diverse backgrounds. They have been involved in projects and played roles in organizations that cross the entire information security industry, ranging across all levels—from practitioner to executive. Additionally, each of the authors has traveled to just about every country in the world speaking to and consulting world governments, financial institutions, retail, defense organizations, utilities firms and more.

These amply-qualified true specialists in their field have produced an important text that succinctly conveys their unique view of the issues dealt with, which I recommend heartily to anyone interested in cyber-crime's inner workings, and particularly to those wishing to formulate ideas for "getting the word out" about cyber-crime and cyber-warfare in general—something the world needs more of.

Don't be just scared—be prepared!

—**Eugene Kapersky**,
CEO and Co-Founder of Kaspersky Lab

Preface

When four loosely associated industry colleagues decided to come together as an author team to write a book, little did we realize how all our combined experience would amount to a hill of beans when it came to in-depth econometric analysis, or how drastically the world of information security would change during the time it took us to write and edit the book. The prophetically named "Project Mayhem" would evolve over time, eventually taking a life of its own, impacting the authors and those around us. We managed to write between overflowing workloads, amazing travel schedules, and disturbingly challenging changes in the information security world. Work and emails changed for more than one of us, and we all learned from each other as we collaborated and progressed in writing the project.

One of the dangers of writing a book that lives on the "bleeding edge" of technology is that you run the risk of simply being proven wrong. As an example, late into the editing stages of this chapter, information regarding the U.S. government's involvement in developing Stuxnet as part of a comprehensive campaign to undermine Iran's nuclear efforts was published as fact by the *International Herald Tribune*. This previously alleged but now published fact could have, if it were missing, made parts of the book substantially less impactful. At the same time, including it without verifiable proof would have caused liability concerns for the authors and the publisher. Your understanding as a reader is greatly appreciated by the authors.

Our combined personal and work experiences, which included front-line law enforcement, pure research, sales, marketing, launching a start-up, surviving a start-up, and more than 40 years of experience in the information security arena allowed us to bring together a very Renaissance approach to the subject.

HOW THIS BOOK IS ORGANIZED

As you read the book, please keep in mind that it is not necessarily intended to be read cover to cover. In fact, it is intended to be an encyclopedic resource that you can reach for time and again as different subjects come across your mind, or your desk.

In Chapter 1, titled "Psychological and Cultural Trends," we address how hacking has seen a significant increase in mainstream media coverage, and how never before has the technical gap between the attackers and victims been so vast. Attack methods take advantage of vulnerabilities that are both so plentiful and so generally invisible as to seem to victims as if they are black magic; if the world were *A Connecticut Yankee In King Arthur's Court*, many of today's cyber victims are unwitting Merlins to the attackers' Hank Morgan. But cyber crime is not a uniform activity—like traditional crime, there are countless variations, motives and methods, and like traditional crime, cybercrime has economic ramifications that transcend the intent of the criminal. This chapter looks at the psychology of the attackers and the victims, and some of the major categories of cybercrime we have seen to date.

In Chapter 2, "Seasons of Change," we explore the ideas that both vex and invigorate criminologists, psychologists, behavioral analysts, and profilers alike by providing a historical perspective going back to the earliest days of phreaking and hacking. At the same time, while we try to reconcile the threats posed by the products and services crafted and delivered by cybercriminals, we investigate how the monetization of cybercrime has influenced the economics of these activities over time.

As we move to Chapter 3, "Drivers and Motives," we cover the advancements in technology that are intended to make the way we communicate better, faster, and more efficient, and how nefarious cybercriminals exploit the use of these new technologies to gather and conduct criminal operations that span the globe. We also explain that the motives that drive specific organizations to commit such crimes vary, but ultimately they revolve around the fact that data has some form of monetary value and it can be monetized.

Chapter 4, "Signal-to-Noise Ratio," deals with a fact that hackers have quickly understood: as their malware becomes more intricate, they must maintain a low "Signal to Noise Ratio" to minimize the chance of getting caught. As such, the most advanced hackers have learned to manage the "command and control" channels of malware as hidden traffic, difficult to pinpoint within the massive amounts of communications that are now part of our everyday hyper-connected lives.

Chapter 5, "Execution," discusses the tools, tactics, techniques, and procedures that cybercriminals commonly use when executing their attacks. In addition to exploring the various forms of seeding, compromise, exploitation (human and system alike), exfiltration, muling, laundering, and expansion that are necessary to ensure successful cybercriminal initiatives, the chapter also takes a look at countermeasures, including counterintelligence, that are useful in detecting, identifying, and arresting the progress of cybercriminals.

Chapter 6, "From Russia with Love," discusses the creation of one of the most alluring and vital sub-economies supporting criminal e-commerce today: the Russian and Eurasian cyber-underground. The historical and geopolitical perspective in this chapter provides an interesting view of a dominant force in the cyber-underworld today.

Chapter 7, "The China Syndrome," provides us a parallel view to the previous chapter on Russia, but it focuses on the Great Cyber-Dragon. It specifically addresses the factors that drive China's role in industrial espionage.

Chapter 8, "Pawns and Mules," investigates what happens after a breach occurs. In particular, it discusses how organizations take advantage of individuals to carry out a cybercrime and how they launder money to stay under the radar. This is a critical chapter in understanding how the monetization of cybercrime has made transformed it from a hobby into an international industry.

Chapter 9, "Globalization," then shows you how the global nature of the Internet and the destruction of geographic barriers that was brought about by the online world have created a Wild West free-for-all for malfeasants, especially in developing countries with limited legal frameworks and questionable enforcement practices. As you will see in this chapter, criminal enterprises have learned that they can use these countries as sources for their attacks with little to no risk, and they can attack government and corporate targets in developed countries with near-absolute impunity.

Chapter 10, "America, Land of Opportunity," addresses how the birthplace of the online community and the largest cyber-economy have created an online eco-system perfect to foster cyber crime. The American digital landscape has created a perfect place for malware to develop and flourish. Given the talent, the freedom, and the target-rich environment, this is definitely not surprising.

Chapter 11, "Global Law Enforcement," shows you how the "good guys" are addressing this economic free-for-all. To this point, we have looked at cyber-criminal activity, and it's reasonable for you to be thinking, "Say, isn't this stuff against the law?" or "What about that 'super-tough cybercrime legislation' my Congressman was raving about last election cycle?" With that in mind, this chapter explores what international and, predominantly, U.S. law enforcement is doing to stanch the flow of money into the criminal underground. What are the laws related to cybercrime and how are they enforced? Which agencies are doing the enforcement and how is prosecution being conducted? How are international agencies cooperating and collaborating to prosecute crimes that cross international boundaries? Most importantly, this chapter describes in detail how a lack of articulated metrics and the definition of cybercrime in the United States has resulted in a turf battle between federal law enforcement

Cyber X: Criminal Syndicates, Nation States, Sub-National Entities and Beyond

INFORMATION IN THIS CHAPTER:

- Classifying the Cyber Actor
- Criminal and Organized Syndicates
- Nation States

INTRODUCTION

The following is a recap of Chapter 7 from *Cyber Crime and Espionage* by John Pirc and Will Gragido. It is a great precursor to the content of the book you now hold in your hands. In the aforementioned chapter, John and Will discuss many concepts and tradecraft that are used by the nefarious cyber actor. This will help readers who are somewhat new to cyber security as it covers several examples of tools, tradecraft, and countermeasures at the very highest level. Additionally, John and Will cover the economic spend of various countries on cybercrime and cyber warfare. The concepts, tradecraft, and economics only scratch the surface in terms of what goes on after a breach or cyber attack occurs. The genesis of Blackhatonomics will go beyond the surface into the details of what happens post-breach by John Pirc, Will Gragido, Daniel Molina and Nick Selby.

The classification and categorization of nefarious cyber actors has moved well past the script kiddie. Fame and bragging rights on compromised systems and Web site defacements are passé and have long ago enjoyed their 15 minutes of fame. It's important to realize that the motive behind the script kiddie or recreational hacker is more ego-driven and bent on destruction of data without a hidden moral, political, or economic agenda. The entities we are about to discuss are motivated by economic, political, and sometimes moral agendas that drive cybercriminals to conduct targeted cyber operations from every corner of

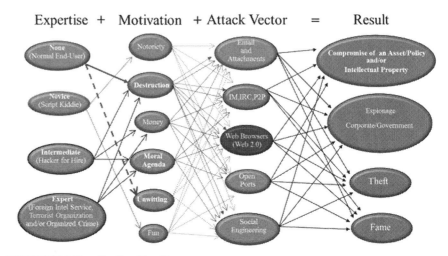

FIGURE I.1 Classification At-a-Glance

the globe. Figure I.1 highlights some key characteristics of today's cyber actor. If you asked 50 different security professionals to classify cyber actors by expertise, motivation, and attack vector, you would get 50 different answers, but in the end, we think we can all agree with some of the factors in the figure. Additionally, the figure really illustrates the multitude of attack vectors that are often leveraged depending on the cyber actor's skill set and expertise.

CLASSIFYING THE CYBER ACTOR

The following subsections describe the categories highlighted in the figure.

Expertise Level

None This is your typical day-to-day end-user. In the eyes of cyber actors, these are like pawns waiting to be compromised with the click of a button. Additionally, they might be patient zero and propagating exploit code without even knowing they have been compromised. The flip side of this is the typical day-to-day end-user gone bad, a once-trusted resource who has become an insider threat with the ability to destroy and exfiltrate critical intellectual property outside the organization.

Novice This is your script kiddie, taking well-known methods of exploitation and hoping the target of the attack is still vulnerable to the exploit. Additionally, the script kiddie ranks right up there with the individuals who perform Web defacements or distributed denial of service (DDoS) attacks for fun or political agendas. In the greater scheme of things, those types of activities are loud, apparent, and easily corrected. That's not to say that experts are not going to use point-and-click, prebuilt, widely distributed attack frameworks. In some rare cases, script kiddie tools have been used to perform certain aspects of what we categorize as an advanced persistent threat. Point-and-click hacking can be found in exploit frameworks that are similar to Metasploit or online Hacking as a Service (HaaS) tools in which individuals can rent/lease botnets and other types of attack tools. Additionally, these individuals might have high-level scripting and coding knowledge.

Intermediate These individuals have very specific skill sets and market themselves in the underground community as providing a wide range of services and abilities for money. There have been cases in which people with these skill sets have performed activities based on moral and religious beliefs. These individuals have experience in writing code in low-level scripting languages, and sometimes have the ability to rewrite or reverse certain aspects of code depending on the target.

Expert These are the most sophisticated cyber actors on the planet. They are typically employed or funded by a foreign intelligence service, national defense organization, organized crime, or terrorist organization. These individuals have the ability to reverse-engineer hardware and software. Additionally, they can write very specific exploit code and encrypt various aspects of the code, and they are fluent in denying attribution through covert channels and darknets to hide their location.

Attack Sophistication Model

The attack sophistication model is a way to determine the abilities of an expert-level adversary. This is important, as the attack sophistication footprint of an expert is far different from that of a novice to intermediate-level cyber actor. We can categorize them into two different tiers.

Tier 2 (Nonkinetic) A great example of this type of sophistication was modeled in what the security industry calls [1]"Operation Aurora." The telemetry of this attack was seen in many of the high-tech compaines out of Silicon Valley. The adversaries that conducted this operation used various known methods to exfilitrate data outside the network. They were able to compromise a critical vulnerability in Microsoft Internet Explorer that led to their ability to conduct the operation. In addition, once they successfully used the browser as their vector to execute their code, the attackers were able to send information about the PC they targeted that included OS, patch information, and so on to a command and control server to provide them with clear insight into other vulnerabilities they could use to harvest whatever data set they wanted to retrieve. Tier 2 attacks are often multistaged attacks that involve multiple vectors, as researchers discovered in Aurora.

---

[1] http://threatpost.com/en_us/blogs/inside-aurora-malware-011910

Advanced Threats

The great thing about Operation Aurora and Stuxnet is that we the security professionals know about them. What is frightening are those classes of attack that are sitting dormant on systems, waiting for a specific instruction set in order to become active. Stuxnet is just one example that was targeted at Siemens gear, but what about other vendors? Additionally, with the rapid outsourcing of engineering and supply chain manufacturing to Third World countries that have very loose controls on those they hire, it might come as no surprise that we might be enabling the delivery of advanced/invisible code in a vendor's product life cycle development process or supply chain insertion. We are not advocating that outsourcing is a bad thing. It makes perfect economic sense in highly competitive markets that require quick time to market and the ability to staff a project with a team of full-time engineers (FTEs) for a fraction of the cost required in their home country. What we are emphasizing is the danger of not knowing the backgrounds of the individuals you're outsourcing source code to, or the contractors that are deploying critical infrastructure. The inside threat is real, and we need to wake up to the realities of the advanced tactics used by adversary countries, crime syndicates, and terrorist organizations in terms of conducting nefarious cyber activities. As we will discuss elsewhere in this book where we cover social engineering and other tactics to gather information, tier 1 players are experts in deception and nefarious cyber tradecraft. Just ask yourself a simple question about Stuxnet: How did someone get malicious code on a closed air-gapped network? There is plenty of speculation in terms of how it was delivered; most people feel it was delivered via USB. However, if it was through a supply chain driven by code that was written internally or inserted during manufacturing, this raises the speculation that this was indeed state-sponsored.

Tier 1 (Kinetic) This type of attack is probably the most sophisticated attack ever written. Finding an example of this type of attack is very difficult, as it is not typically shared within the general security community. These attacks are usually targeted at air-gapped networks or networks that would be considered highly secured, such as power companies (SCADA networks), governments, and defense organizations. Additionally, they require deep insight into a specific vendor's code base and product offering. These attacks can evolve into kinetic-based attacks. In 2007, the Idaho National Laboratory conducted a project called, oddly enough, [2]"Aurora Test." For this project, the lab created about 21 lines of code and injected it into a closed-test SCADA network that caused a generator to blow up. This project proved that the ability to weaponize code and use it to conduct kinetic activities is a sad reality in terms of threat landscape maturity. What's even more alarming about the weaponization of malicious code is that such code could end up in the hands of a terrorist organization. A timely example of a tier 1 attack is Stuxnet. At the time of this writing, there is no known patch to fix this very sophisticated attack. The attack was so targeted that it went after a

---

[2] http://www.globe-expert.eu/quixplorer/filestorage/Interfocus/6-Science_Technologie/66-Aeronautique_Espace/66-SRCNL-AviationWeek_com_Homepage/201009/CyberAttack_Turns_Physical.html

piece of SCADA gear developed by Siemens—in particular, two Siemens program logic controllers (PLCs). Additionally, there was a lot of intelligence wrapped in the code: In fact, it was smart enough to discern what devices to use to arm its destructive payload, and it had the ability to terminate after a predefined date. What's important to note is that tier 1 attacks do require a "pawn" to deliver the malware, as in the case of Stuxnet, these types of infrastructures are air-gapped.

Advanced Persistent Threats

Tier 2 and tier 1 attacks can be categorized under the umbrella of advanced persistent threats, due to their level of severity and sophistication. Although advanced persistent threats are not new, they received a lot of media attention in 2010 with Operation Aurora and Stuxnet, and as a result, the broader security community is only beginning to get a taste of their maturity and sophistication.

Modus Operandi

The great thing about cybercrime, both state-sponsored and non-state-sponsored, is that sometimes the criminals use the same modus operandi in terms of malware and command and control nodes on the Internet, although these command and control nodes can go online very quickly and just as quickly can be brought down. However, companies such as Damballa, which is leading the industry in botnet detection and remediation, have found similarities in criminal activity from various nefarious cyber actors. Based on the type of malware and command and control infrastructures involved, they can assign group names that help them identify similarities that are carried out by nefarious cyber actors.

In terms of the attack sophistication model, this would apply to tier 2 and some tier 1 attacks. Tier 1 attacks often involve malware that might not call back or beacon to the Internet, as these attacks are typically on air-gapped networks. In specific cases such as Stuxnet, researchers were able to find clues left by the author of the code. For example, researchers found the numeric string 19790509 in the Stuxnet code, which is ISO 8601 for capturing dates. According to *Wired* magazine, [3]"Researchers suggest this refers to a date—May 9, 1979—that marks the day Habib Elghanian, a Persian Jew, was executed in Tehran and prompted a mass exodus of Jews from that Islamic country." Additional messages were found in the code that indicated that it came from Israel, or the United States because of its support of Israel. Additionally, extremist groups such as terrorists are big on dates and conducting operations that coincide with those dates. Our thoughts on the matter might differ: Deception is key, and someone could have easily placed those markers in the code. That date

---

[3] www.wired.com/threatlevel/2010/10/stuxnet-deconstructed/#ixzz11Eh1Y5vc

is also the anniversary of the second Unabomber attack. Does this mean that the code was created at Northwestern University? At the time of this writing, we have not come across anything that links attribution or modus operandi to a state-sponsored actor. However, the sophistication of this specific piece of malware and its possible destructive properties indicates that it's highly suspected that a criminal organization didn't create it. Modus operandi is just an additional step on our way toward attribution.

The Importance of Attribution

Every advanced attack that is highly publicized today seems to point the finger of origin back to China. As security professionals, we would love to believe that attribution was so simple. We've come across situations were the geo-location of an IP address was mapped back to a specific province in China. But imagine what would happen if a defense agency wanted to respond with kinetic means to a cyber attack, and found out it just launched an attack on unwitting individuals. That is why attribution is so important, and to just lay the blame on China is becoming more of an annoyance.

Figure I.2 shows how easy it can be to trace an IP address. If we were to use a TOR client or anonymous proxy and run the same lookup, we would receive an entirely different result. Additionally, even if attribution can be traced back to a source country, it does not necessarily mean that it's state-sponsored; it could be a few bored college students having fun. That's why it's important to look at the attack sophistication model, modus operandi, and origin of the attack. The following is data for tracing attribution based on IP address. This happens to be the geo-location of one of the authors. The city and postal codes are incorrect; however, if someone with authority contacted the ISP with the IP address and host name, he or she would easily be able to trace this back to one of the authors.

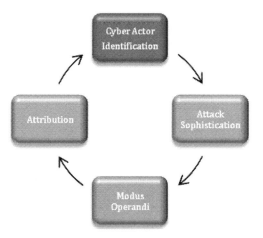

FIGURE I.2 Cyber Actor Identification

www.maxmind.com/app/ip-location
Your main IP address: X.X.X.69 United States
Location (from MaxMind database) city: Cedar Park
Region name: Texas
Latitude: 30.4998
Longitude: -97.8082
Postal code: 78613
Local IP addresses detected: 10.0.1.5
Browser variables that may reveal your system, time zone, and language:

> Date: Sun Oct 03 2010 10:35:48 GMT-0500 (CDT)
> User Agent: Mozilla/5.0 (Macintosh; U; Intel Mac OS X 10_6_4; en-us)
> AppleWebKit/533.18.1 (KHTML, like Gecko) Version/5.0.2
> Safari/533.18.5

Standard HTTP request variables that may reveal your system, language, or indicate proxy usage:
HTTP_ACCEPT_CHARSET
HTTP_ACCEPT_ENCODING: gzip, deflate
HTTP_ACCEPT_LANGUAGE: en-us
HTTP_CACHE_CONTROL:
HTTP_CONNECTION: keep-alive
HTTP_USER_AGENT: Mozilla/5.0 (Macintosh; U; Intel Mac OS X 10_6_4; en-us) AppleWebKit/533.18.1 (KHTML, like Gecko) Version/5.0.2 Safari/533.18.5

We ran another test using a proxy anonymizer and were traced back to the United Kingdom. That is why attribution is so important. Uncovering the real IP address and geo-location of someone is not that easy and sometimes requires working with national and international ISPs.

Your main IP address: X.X.X.130 United Kingdom
Location city: London
Region name:
Latitude:
Longitude:
Postal code:
Local IP addresses detected: 10.0.1.5
Browser variables that may reveal your system, time zone, and language:

> Date: Sun Oct 03 2010 10:50:48 GMT-0500 (CDT)
> User Agent: Mozilla/5.0 (Macintosh; U; Intel Mac OS X 10_6_4; en-us)
> AppleWebKit/533.18.1 (KHTML, like Gecko) Version/5.0.2
> Safari/533.18.5

Standard HTTP request variables that may reveal your system, language, or indicate proxy usage:
HTTP_ACCEPT_CHARSET
HTTP_ACCEPT_ENCODING: gzip, deflate
HTTP_ACCEPT_LANGUAGE: en-us
HTTP_CACHE_CONTROL:
HTTP_CONNECTION: keep-alive
HTTP_USER_AGENT: Mozilla/5.0 (Macintosh; U; Intel Mac OS X 10_6_4; en-us) AppleWebKit/533.18.1 (KHTML, like Gecko) Version/5.0.2 Safari/533.18.5

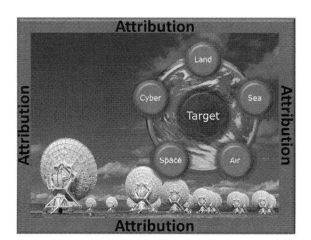

Additionally, if you wanted to hide your tracks on the Internet, it's not that hard. A nefarious cyber actor can launch an attack from China and a lot of different anonymous connectors (onion routed networks, proxies, or darknets), making the attack look like it's coming from Austin, Texas, or Chicago.

Now that we have touched on the finer points of classification, attack sophistication, modus operandi, and attribution characteristics associated with categorizing the entities responsible for the cyber activities we read about in the media, it's time to talk about criminal and organized syndicates.

CRIMINAL AND ORGANIZED SYNDICATES

Cybercrime has become so profitable for criminals that it has surpassed the drug trafficking trade, according to recent reports from the FBI. It's much easier for a criminal to conduct nefarious activities online than to physically break into a bank or someone's home. A recent article posted on net-security. org provides great details on the dynamics of cyber mafia-like activities on the Internet. The following are the roles that are played out in these types of organizations:

1) [4]*"The coder, the 'techie' (that keeps the servers and ISPs online),*
2) *The hacker (actively searches for vulnerabilities to exploit),*
3) *The money mule, the fraudster (creates social engineering schemes), and others"*

The money mule is an important aspect of the operation. The money mule is the one in charge of actually setting up multiple bank accounts with multiple false identities. A great example of this is the recent ZeuS Trojan bust that was targeted at the banking industry. In this case, Ilya Karasev of Russia entered the United States on a J-1 visa and then converted his status to an F-1 student visa. Under this type of visa, a foreigner has the right to open a bank account in the United States. However, Karasev opened three accounts, under three different passports, all with the same bank but at different branch offices. In order to fly under the radar and avoid being reported to the IRS, Karasev never exceeded the amount of $10,000 in each wire transfer or deposit. What's alarming about this case is that this criminal possessed multiple passports from different countries and used them effectively within the United States. As you can see, cybercriminals are very good at their tradecraft and are willing to risk a lot for what might be a significant payout in the end.

The Karasev case is an example of a recent case that has ties to Eastern Europe cybercrime rings. The first organization that comes to mind when talking about cybercrime syndicates is the Russian Business Network (RBN). RBN has allegedly been tied to the Storm botnet and MPack, a pay-for-hacking tool that can run from $500 to $1,000. Most cybercriminals specialize in identity theft, stolen credit cards, money laundering, framework exploitation, and selling services that enable other cyber actors to rent/lease botnets and other nefarious services.

Of course, Russia is not the only country that allegedly has cybercrime rings running within its borders. Cybercrime has also been traced to China and an organization called Honker Union of China. It's reported that this organization, best known for its attack on the White House's Web site, has about 80,000 members and is vocal in communicating its activity. For example, it recently published the following calling on its Web site (see Figure I.3).

Translated into English, the Web site reads as follows:

[5]*"Notice to Honker Union general members!*

Recently, tension has been built up between China and Japan, some of the patriotic hackers and honkers also are ready to make a move, boldly publicizing to launch

---

[4] www.net-security.org/secworld.php?id=9060
[5] www.chinahush.com/2010/09/15/honker-union-of-china-to-launch-network-attack-against-japan-is-a-rumor/

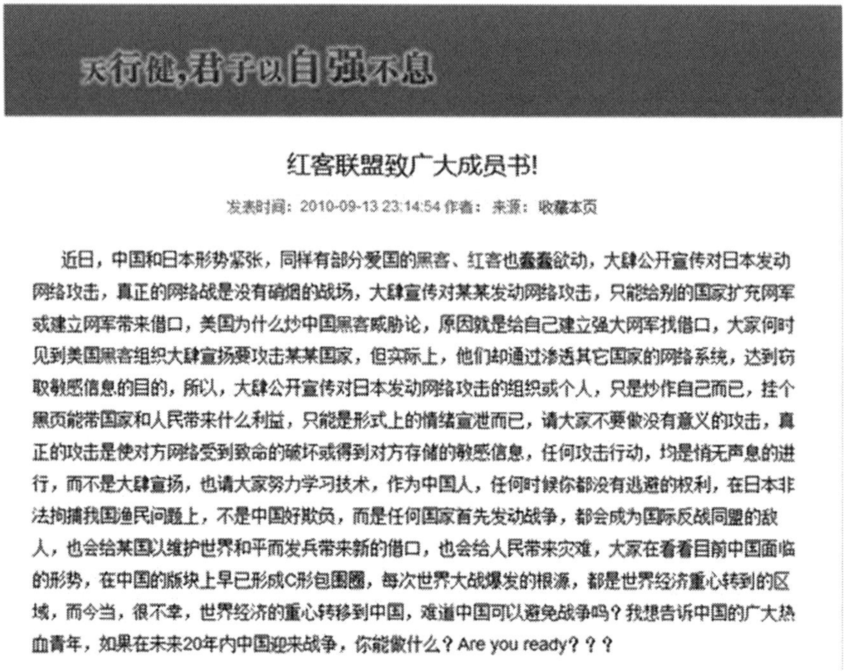

天行健,君子以自强不息

红客联盟致广大成员书!

发表时间: 2010-09-13 23:14:54 作者: 来源: 收藏本页

近日，中国和日本形势紧张，同样有部分爱国的黑客、红客也蠢蠢欲动，大肆公开宣传对日本发动网络攻击，真正的网络战是没有硝烟的战场，大肆宣传对某某发动网络攻击，只能给别的国家扩充网军或建立网军带来借口，美国为什么炒中国黑客威胁论，原因就是给自己建立强大网军找借口，大家何时见到美国黑客组织大肆宣扬要攻击某某国家，但实际上，他们却通过渗透其它国家的网络系统，达到窃取敏感信息的目的，所以，大肆公开宣传对日本发动网络攻击的组织或个人，只是炒作自己而已，挂个黑页能带国家和人民带来什么利益，只能是形式上的情绪宣泄而已，请大家不要做没有意义的攻击，真正的攻击是使对方网络受到致命的破坏或得到对方存储的敏感信息，任何攻击行动，均是悄无声息的进行，而不是大肆宣扬，也请大家努力学习技术，作为中国人，任何时候你都没有逃避的权利，在日本非法拘捕我国渔民问题上，不是中国好欺负，而是任何国家首先发动战争，都会成为国际反战同盟的敌人，也会给某国以维护世界和平而发兵带来新的借口，也会给人民带来灾难，大家在看看目前中国面临的形势，在中国的版块上早已形成C形包围圈，每次世界大战爆发的根源，都是世界经济重心转移的区域，而今当，很不幸，世界经济的重心转移到中国，难道中国可以避免战争吗？我想告诉中国的广大热血青年，如果在未来20年内中国迎来战争，你能做什么？Are you ready？？？

FIGURE I.3 Original "Notice to Honker Union general members"

network attacks on Japan. The real war on the networks has no smoke and fire. Publicizing to launching cyber attacks against certain country can only give excuses for other country to establish network army and network forces. Why does the United States claim Chinese hackers a threat? The reason is to give excuses for themselves to build up a strong network army. When have you ever heard the American hackers organizing publicly to launch cyber attack against certain country? But in fact, they meet the objective of stealing sensitive information by infiltrating other countries' network systems. Therefore, the organization or the person who boldly publicized to launch network attacks against Japan is only doing a publicity stunt for themselves. What benefit can hacking a web page bring our country and the people? It is only a form of emotional catharsis, please do not launch any pointless attacks, the real attack is to fatally damage their network or gain access to their sensitive information. Any attack will be executed silently, rather than vigorously promoting it. And also everyone please work hard on learning technologies, as Chinese, you have no right to escape the responsibility at any time. On the issue of Japan illegally arrested our fishermen, it is not that China is easy to be bullied, but any country that starts a war will become the enemy of the international anti-war alliance, which will give certain country new excuses to send troops to maintain peace in the world, and also will bring disaster to the people. Please take a look at the situation China is facing today, China on the map is already being surrounded by a c-shaped ring. Every world war always broke out

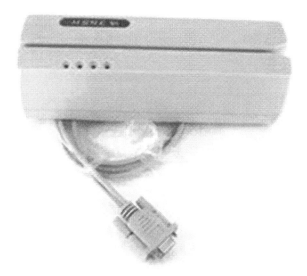

FIGURE I.4 MSR 206

from where the world economy shifted to, and today, unfortunately, the world economy center is shifting to China, can China avoid a war? I want to tell the vast number of passionate young people in China, if China is in war in the next 20 years, what can you do? Are you ready???".

Although this type of messaging goes against the modus operandi of typical organized crime groups, the Honker Union of China boasts more than 80,000 members who can carry out nefarious activities.

Another group in China, called Black Hawk, was shut down by Chinese authorities for selling exploit tools and teaching the tradecraft associated with hacking. It's been reported that Black Hawk made more than $1 million and had more than 12,000 paying members.

Other tools that cybercriminals use are magnetic stripe readers or writers (see Figure I.4). These enable cybercriminals to populate and read data from credit cards and other media that use magnetic stripe readers.

Another technique that is commonly used, but requires physical interaction with the target, is ATM skimming. An ATM skimmer blends right in with an ATM, and at first glance you might find it difficult to identify the device. The skimmer mounts directly over the slit into which you insert your credit card (see Figure I.5). Additionally, it often has a pinhole camera that takes a picture when you enter your PIN on the ATM.

As we mentioned, the majority of cybercrime is conducted in a logical manner, with the exception of ATM skimming, which requires that you physically

FIGURE I.5 ATM Skimmer Device

deploy and harvest once the operation has been conducted. As you can see, cybercrime is a major, lucrative business that is fueled by the almighty dollar and the ability to cash in on the lowest common denominator in terms of attack vector.

NATION STATES

Nation states have often been the focus of major Internet attacks that have been targeted at nation state networks and Web servers. Unlike the cybercriminal trying to turn a buck or make money from his or her nefarious cyber operations, nation states have an entirely different agenda. Operations that are run from nation states can range from disinformation to economic, political, and/ or military gain. The great thing about nation state cyber activities is that they are well funded, employ some of the world's most talented security engineers, and for the most part are under a veil of secrecy. Nation state cyber programs often operate under the direction of the country's defense organizations, foreign intelligence services, and law enforcement services. Additionally, some nation states have been known to fund subnational entities such as terrorist and extremist groups.

Subnational Entities

A great example of a terrorist group that is state-sponsored is Hezbollah, which operates out of Lebanon and received a lot of its military and tactical training from Iran's Revolutionary Guard. Although Hezbollah is seen as a positive enabler to the social services fabric in the eyes of the Lebanese, it is deemed a

research and development into [7]"network-based combat, including cyber-espionage and counter-espionage". The Chinese military has what is known as a [8]"Military Intelligence Department" that contains seven different bureaus, each responsible for a specific task. One of the bureaus deals with cyber intelligence operations for conducting espionage, surveillance, and other electronic means for gathering intelligence. In addition to China's government cyber program, they also are integrated with their countries' major universities and research and development organizations. Based on the shear size and population of China and its aggressive stance in expanding its cyber operations, it's likely that China will continue to be one of the key players in this area. The following is an example of China's capabilities as of May 2008:

[9]*"China People's Liberation Army (PLA) Military Budget: $62 billion* ☐

Global Rating in Cyber Capabilities: #2

Cyber Warfare Budget: $55 million

Offensive Cyber Capabilities: 4.2 (1 = Low, 3 = Moderate, and 5 = Significant)

Cyber Weapons Arsenal (In Order of Threat):

> *Large, advanced botnet for DDoS and espionage*
> *Electromagnetic pulse weapons (non-nuclear)*
> *Compromised counterfeit computer hardware* ☐
> *Compromised computer peripheral devices* ☐
> *Compromised counterfeit computer software* ☐
> *Zero-day exploitation development framework* ☐
> *Advanced dynamic exploitation capabilities* ☐
> *Wireless data communications jammers* ☐
> *Computer viruses and worms* ☐
> *Cyber data collection exploits* ☐
> *Computer and network reconnaissance tools* ☐
> *Embedded Trojan time bombs (suspected)* ☐
> *Compromised microprocessors and other chips (suspected)* ☐

Cyber Weapons Capabilities Rating: Advanced ☐

Cyber Force Size: 10,000 + ☐

Broadband Connections: More than 55 million"

- **Russia** This country possesses a mature cyber warfare model and doctrine. This was very evident during the recent altercation between Russia and Estonia. The capabilities demonstrated during that cyber campaign basically took the entire country of Estonia off the Internet grid. The following

---

[7] www.atimes.com/atimes/China/LB09Ad01.html
[8] www.atimes.com/atimes/China/LB09Ad01.html
[9] http://defensetech.org/2008/05/08/chinas-cyber-forces/1

is a brief synopsis from Kevin Coleman on the cyber capabilities that Russia is known to have as of May 2008:

[10]*"Russia's 5th-Dimension Cyber Army:*

Military Budget: $40 billion

Global Rating in Cyber Capabilities: Tied at # 4

Cyber Warfare Budget: $127 million

Offensive Cyber Capabilities: 4.1 (1 = Low, 3 = Moderate, and 5 = Significant)

Cyber Weapons Arsenal (In Order of Threat):

- *Large, advanced botnet for DDoS and espionage*
- *Electromagnetic pulse weapons (non-nuclear)*
- *Compromised counterfeit computer software*
- *Advanced dynamic exploitation capabilities*
- *Wireless data communication jammers*
- *Cyber logic bombs, computer viruses, and worms*
- *Cyber data collection exploits*
- *Computer and network reconnaissance tools*
- *Embedded Trojan time bombs (suspected)*

Cyber Weapons Capabilities Rating: Advanced

Cyber Force Size: 7,300 +

Reserves and Militia: None

Broadband Connections: 23.8 million +"

The bottom line with Russia is that it is very advanced in terms of information operations, and like China, it has many universities from which to pick and choose engineers. According to an article by Kevin Coleman, Russia graduates more than 200,000 people in science and technology every year. That's not to say all will join the government, but this gives Russia an extremely large talent pool from which to select highly qualified individuals.

■ [11]*"**Iran** The following is a brief example of the cyber capabilities that Iran possesses.

Estimated Cyber Capabilities

Iran Islamic Revolution Guards Corps (IRGC)

- *Military Budget: $11.5 billion*
- *Global Rating in Cyber Capabilities: Top 5*

---

[10] http://defensetech.org/2008/05/27/russias-cyber-forces/
[11] http://defensetech.org/2008/09/23/iranian-cyber-warfare-threat-assessment/

- *Cyber Warfare Budget: $76 million*
- *Offensive Cyber Capabilities: 4.0 (1 = Low, 3 = Moderate, and 5 = Significant)*

Cyber Weapons Arsenal (In Order of Threat):
1. *Electromagnetic pulse weapons (non-nuclear)*
2. *Compromised counterfeit computer software*
3. *Wireless data communication jammers*
4. *Computer viruses and worms*
5. *Cyber data collection exploits*
6. *Computer and network reconnaissance tools*
7. *Embedded Trojan time bombs (suspected)*

Cyber Weapons Capabilities Rating: Moderate to advanced
Cyber Force Size: 2,400

- *Reserves and Militia: Reserve with an estimated at 1,200*
- *Broadband Connections: Less than 100,000"*

These are just a few examples of the capabilities that nation states have in their cyber arsenals. The United States, United Kingdom, France, India, Pakistan, North Korea, and Japan have very mature cyber warfare models and doctrines that provide them with very specific capabilities to carry out various levels of cyber operations.

SUMMARY

In this preface, we discussed the capabilities of various cyber actors and models that help articulate the characteristics and sophistication levels of a variety of groups. One key element is attribution of the attacker, and toward that end, we gave a few examples of methods for tracing attribution. Attribution on a global level does require a lot more analysis and clarity, however. In terms of criminal activity across borders, it requires the help of state and local law enforcement and information from ISPs, which can take a long time if the attack is coming from another country. The cyber actors that are involved in cybercrime, cyber warfare, and cyber terrorism are driven by economic, political, and moral agendas. We've seen the threat landscape constantly evolving over the past two decades. These changes have shaped the dynamics of what we are dealing with today in terms of the threat landscape. Here is a brief walk down memory lane and a glimpse at what the future will hold if we continue at this pace:

- **The first decade (1990–1999)** The Internet was a nice-to-have or a luxury. The profile of the attacker was all about control and named individuals taking responsibility for Web defacements, worm propagation, and so on.
- **The second decade (2000–2009)** The Internet was a necessary tool for competing on a global level and staying connected from a personal

perspective. This era presented us with many challenges, as e-commerce, online banking, and other technological advances provided the nefarious cyber actor with many targets that he or she could attack for financial gain. Additionally, nation states began to regard the Internet as a national asset, and as we mentioned, began spending upward of $1 billion in order to defend it.

- **The third decade (2010–Present)** As we move into a new decade and threat paradigm, it's likely that we will witness a cyber kinetic attack. Stuxnet was a great example of what could have been a successful cyber kinetic attack. In the event that a cyber kinetic attack does occur, the attribution factor might be hard to prove, but from what we have learned in terms of terrorist organizations, they are the only ones that will claim publicly that they were responsible for such an attack. At least this gives the analyst and security experts working the case a place to start. With each new decade and major technology innovation driving us closer toward a society that depends on being connected, the attack landscape will only become wider and much harder to defend if we give security a back seat or treat it as a checkbox.

References

[1] Coleman, K. (September 15, 2010) "Honker Union of China to launch network attacks against Japan is a rumor". China Hush. Retrieved October 5, 2010, from www.chinahush.com/2010/09/15/honker-union-of-china-to-launch-network-attack-against-japan-is-a-rumor/.

[2] Coleman, K. (June 2, 2008) Hezbollahâ€™s Cyber Warfare Program | Defense Tech. Defense Tech | The future of the Military, Law Enforcement and National Security. Retrieved October 5, 2010, from http://defensetech.org/2008/06/02/hezbollahs-cyber-warfare-program/.

[3] Dumitrescu, O. (July 16, 2010) "Considerations about the Chinese Intelligence Services (II)". Conflict Resolutions and World Security Solutions. Retrieved October 5, 2010, from www.worldsecuritynetwork.com/showArticle3.cfm?article_id=18347&topicID=66.

[4] Fisher, D. (n.d.). "Inside The Aurora (Google Attack) Malware." The First Stop for Security News. Retrieved October 5, 2010, from http://threatpost.com/en_us/blogs/inside-aurora-malware-011910.

[5] Fulghum, D. (October 5, 2010) "Cyber-Attack Turns Physical". Aviation Week. Retrieved September 28, 2010, from www.aviationweek.com/aw/generic/story_channel.jsp?channel=defense&id=news/asd/2010/09/27/05.xml&headline=Cyber-Attack%20Turns%20Physical.

[6] Krebs, B. (January 15, 2010) "Would You Have Spotted the Fraud?" Krebs on Security. Krebs on Security. Retrieved October 5, 2010, from http://krebsonsecurity.com/2010/01/would-you-have-spotted-the-fraud/.

[7] Lam, W. (February 9, 2010) "Beijing beefs up cyber-warfare capacity". Asia Times Online. Retrieved October 5, 2010, from www.atimes.com/atimes/China/LB09Ad01.html.

[8] Langner, R. (October 4, 2010) Stuxnet logbook. Langner. Retrieved January 5, 2010, from www.langner.com/en/.

[9] Macartney, J. (February 9, 2010) "Chinese police arrest six as hacker training website is closed down". Times Online. The Times | UK News, World News and Opinion. Retrieved October 5, 2010, from www.timesonline.co.uk/tol/news/world/asia/article7019850.ece.

[10] Staff, C. (July 15, 2010) Hezbollah (a.k.a. Hizbollah, Hizbu'llah) – Council on Foreign Relations. Council on Foreign Relations. Retrieved October 5, 2010, from www.cfr.org/publication/9155/hezbollah_aka_hizbollah_hizbullah.html.

[11] Vijayan, J. (September 30, 2010) "Zeus Trojan bust reveals sophisticated 'money mules' operation in U.S". Computerworld. Computerworld - IT news, features, blogs, tech reviews, career advice. Retrieved October 5, 2010, from www.computerworld.com/s/article/9189038/Zeus_Trojan_bust_reveals_sophisticated_money_mules_operation_in_U.S.?taxonomyId=82&pageNumber=2.

[12] Villeneuve, N. (October 5, 2010) Vietnam & Aurora. nartv.org. Retrieved April 5, 2010, from www.nartv.org/2010/04/05/vietnam-aurora/.

[13] Zetter, K. (October 1, 2010) New Clues Point to Israel as Author of Blockbuster Worm, Or Not | Threat Level | Wired.com. Wired News. Retrieved October 5, 2010, from www.wired.com/threatlevel/2010/10/stuxnet-deconstructed/#ixzz11Eh1Y5vc.

[14] Zorz, Z. (March 24, 2010) The rise of Mafia-like cyber crime syndicates. Help Net Security. Retrieved October 5, 2010, from www.net-security.org/secworld.php?id=9060.

Psychological and Cultural Trends

INTRODUCTION

When the average nontechnical person reads the newspaper and sees stories about Chinese hackers launching cyber espionage attacks against U.S. chemical companies, the whole thing sounds, frankly, a little *Mission: Impossible*. As they extend their arms and rapidly curl their index and middle fingers while they say the word *spies* or *espionage*, we can almost hear the air-quoted wink as Fortune 500 executives discuss the subject.

The second decade of the 21st century has seen rapid and highly disruptive technical innovation. However, the reason for the prevalence and success of cybercrime is not technical, but rather psychological and cultural: Generally speaking, we have not adapted quickly enough to see (let alone believe) the vulnerabilities that have been created by our intense reliance on the Internet and our constant connectivity to it.

Criminals, though, as they have historically, have quickly adapted to the new and improved Web speed of crime.

At the same time, we have observed that the gulf between the mindset of the attackers and the mindset of the victims symbiotically creates a perfect storm, which is peculiar to this specific moment in history. Never before have the speed of technological advancement, relative slowness in crafting and adopting new legislation, and psychology of criminals and victims combined to create an atmosphere that so encourages and rewards an illegal activity.

In this chapter, we'll examine these vulnerabilities, and the cultural and psychological barriers that prevent us as a society from taking more serious action. This is probably the least technical chapter in this book, but it sets the stage for the cyber attackers we describe later to enter our lives and our companies, and to so successfully relieve us of the intellectual property which, until recently, created the barrier to competing with Western, specifically American, high-technology firms.

PSYCHOLOGY OF ATTACKERS

We can think of few criminal enterprises in which the risks are so low and the potential rewards are so high than that of cybercrime. In this book, when we speak of hackers we are speaking of professional criminal hackers, or those hired by them and acting on their behalf.

Some Background on Cybercrime Legislation

It's a great time to be a cybercriminal: Not only have the laws of most countries not yet caught up with the technology (let alone the crime), but the politics of creating cybercrime laws are mired in a power struggle between agencies in single countries, and are stuck in an absolute gridlock when more than one country is involved. For the past several years, the FBI has struggled in turf wars with other federal, state, and local agencies to reign dominant in the investigation and prosecution of cybercrimes, while other, arguably more capable and proactively talented agencies, such as the United States Secret Service and U.S. Marshals Service (and some agencies which might be simply more contextually appropriate, such as the U.S. Postal Service), are left to fight for table scraps at the budgetary banquet. Simply put, no lawmaker understands this stuff enough to argue very effectively for or against anything yet.

Lastly, it still just isn't very sexy to sponsor cybercrime legislation. Constituents do not yet have the situational awareness necessary to rally behind it, let alone demand it, or they are too caught up in fixing physical infrastructure problems to care much about this "exotic" and seemingly remote problem: To them, cybercrime is the stuff of movies, or something that happens to someone else.

Even a cursory glance through proposals over the past couple of years to strengthen cybercrime law [1] reveals a range of ineffectual options: from the overly broad and relatively meaningless National Security Council Strategy to Combat Transnational Organized Crime [2] to congressional folks of one flavor or another baying for "tougher" "cybercrime" "Legislation".[3] For the most part, these proposals fall into the knee-jerk category of "Oh, *crud*, some of my constituents got cyber-robbed and I had better get something *done*, dammit." This means we get some real whirligig doozies of cyber stinkers, usually centered on the completely false premise that lengthening sentences for computer intrusions [4] is worth doing. It is not. There are laws against hacking,[1] and they come with stiff prison sentences. The problem is not the deterrent nature of the prison sentence, but simplifying the process of establishing the facts of a cybercrime case, articulating the crime and the accompanying mental state of the perpetrator to a jury, and getting the jury and the judge to understand that (a) a crime took place and (b) that guy in the defense dock did it—provided anyone could identify the defendant and that the jurisdictional fruit salad cooperated enough for him[2] to be sitting in court.

No, the problem is not that the sentences are insufficiently severe. The problem is that no cops other than a small number of feds are empowered, prepared, and trained to investigate cybercrime. These numbers are so small that simple resource-based triage means less than 0.01 percent of cybercrimes are even investigated, let alone prosecuted.

Cybercrime legislation, therefore, is not being driven by demands by judges and juries and prosecutors and cops and city officials and stakeholders for better clarity into the issues and better tools with which to do the job. It is being driven by chest-pounding lawmakers seeking to "do something" about the problem.

Enter the Hackers

Against this backdrop, and keenly aware of their unique moment in history, are gangs of professional cybercriminals, most commonly referred to as "hackers," and state-sponsored entities whose sole mission is to disrupt the commercial infrastructures of enemy countries. Previous books on hackers and hacking tended to get weighed down by the personality traits of hackers—depicting mainly male, acne-faced teens and young adults dressed in black and perpetrating their crimes in black-lit rooms of various types. This has long ceased to be the case; in fact, it is a cliché to say that the days of sport hacking and attraction to hacking's seductive subculture have ended, [5] replaced by an industry that exploits computer application vulnerabilities to allow establishment of presence on a network for the purpose of stealing intellectual property.

So, in this book, we're going to talk about the hackers who typically face large corporations. They are well financed, are organized, and have either analyzed

---

[1] Well, there are lots of state and federal laws against unauthorized intrusion into computer networks, which is the same thing.
[2] It's usually a "him," but of course, this could also and just as easily be a "her."

the salability of the information they pilfer or are controlled by a government-sponsored group or organization. In fact, as we will discuss later, from the victims' standpoint it really doesn't matter whether the attackers are government, private sector, independent, or affiliated:[3] They are among the group of people who understand, as David Etue so succinctly put it, that $10 million spent on hacking that steals $1 billion of R&D is a good deal.

Hacking has become the shortest distance between the intellectual property assets you have and those you want, and whether your hacker seeks glory, political advantage, philosophical or religious statements, or cold, hard cash, the psychology of today's professional hacker is merely that of the pragmatist.

They will never send in the A-team if the B-team or C-team can do the job less expensively and as effectively. They will never mount a single campaign when two or three or more can be launched simultaneously. They will never use a previously unseen attack if an oldie-but-goodie gets the job done. In fact, they will always seek the simplest undetectable attack, and then move to quickly understand and then totally dominate the target environment, until they have extracted their quarry and can leave the network. They prefer to do this undetected, but aside from some tactics, being detected by the victim is not a game-changer.

By dominating the Dynamic OODA loop [6] of their victims, the attackers can play endless rounds of whack-a-mole at a very low cost, all the while understanding the cost to their victims in treasure, patience, stress, and professional relationships. Attackers take advantage of the fact that they often understand the playing field—that is, the network that is under attack—better than its owner. In fact, most contemporary and sophisticated attacks rely on the stability of the network to turn single attacks into data-theft endeavors that are long lasting and profitable for the attackers.

Since total dominance followed by exfiltration of the desired data is the goal, prior methodologies of understanding hacker motivations should be superseded with the concept that, if you're determined (for political, philosophical, theological, or financial reasons) to turn to crime, there's plenty of encouragement to make yours a cybercrime.

A former Microsoft employee and former FBI agent once stated it best: "If you commit a cybercrime, there's almost no chance you're going to be caught. If you are caught, there's almost no chance you're going to be prosecuted. If you are prosecuted, there's almost no chance you're going to be convicted. If you are convicted, there's almost no chance you'll serve the full sentence."

---

[3] From the defenders' standpoint, the nature of your advanced adversary doesn't matter at all. See Greg Hoglund's blog about this at http://fasthorizon.blogspot.com/2011/09/apt-plain-hard-truth.html.

Police-Led Intelligence's[4] Dave Henderson, a 15-year veteran police officer, cyber investigator and fugitive hunter, has said it even more succinctly: "If you're a reasonably intelligent criminal, you do the math. You can knock over a 7-Eleven or a bank [and] net three grand and a really good shot at an aggravated felony charge, or you can commit a cybercrime, net 100 times that, and if you're caught, stand a real good chance of doing no time whatsoever—because the cops aren't going to understand what happened and the feds are going to triage your crime out of their workflow." Throw in a single international hop into your attack, and the odds of capture diminish logarithmically toward zero.

If all that is true—and as investigators, incident response consultants, and police officers, we aver that it is—there's almost no reason for any self-respecting, reasonably intelligent criminal *not* to resort to cybercrime.

In addition, and this is the most important point to understand in this section on the psychology of the attacker, there's no reason for your attacker to go anything less than full bore. Armed with the knowledge that they are effectively immune from prosecution, professional cybercriminals are bold, audacious, relentless, remorseless, and utterly devoid of sympathy for their victims.

PSYCHOLOGY OF VICTIMS

On the other side of the chessboard sit the victims, who are as keenly unaware of their moment in history as the attackers are aware of it. Because so many of the most disruptive advances in technologies available to users have occurred on the server side, or back end, of the user experience, to users, detecting the full implications of these revolutionary technological changes is very difficult.

Consider, for example, that to the typical user of technology in a large enterprise, the entirety of the user experience is done through a Web browser, Microsoft Office, Outlook e-mail, and the occasional internal application. To this user, the fact that the browser is the gateway to a world of synchronous backup and server-side magic is totally invisible—and this is exactly the way it is supposed to be! And because most of the interactive work completed by this typical enterprise user consists of invisibly accessing massive stores of data, the user is almost entirely unaware of the power that his or her little terminal might afford an attacker.

---

[4] http://policeledintelligence.com, run by Dave Henderson and Nick Selby, an author of this book, is an advertiser-free Web site dedicated to issues of law enforcement technology and intelligence.

Add to this the fact that most people want to be helpful, [7] so most people won't believe they're the victims of a crime they can't see, hear, feel, smell, or touch; rather, such a person is more likely to say of his or her computer, "I don't have anything on it worth stealing anyway!" [8] As a result, these people are almost the perfect facilitators of cybercrime.

You've often heard it stated that the users are a network's largest single point of vulnerability, but until you contemplate just how easy it is to get users to betray your network security, that statement probably doesn't hit home. Perhaps the best example of this in the past decade came from a UK study in 2004, in which more than 70% of users were not just willing but actually did reveal their computer password to a man-in-the-street interviewer in exchange for some chocolate.[9] Lest this appear to be an outlier, a second, separate study conducted right around the same time showed that nearly 8 in 10 would happily hand over information such as their date of birth, mother's maiden name, and other information necessary to steal their identity.[10]

We will look at each of these common human vulnerabilities shortly, but first, a word or two about the "novelty" of cybercrime.

It's Not the Crimes That Are New, It's Their Execution

One of the best home-team advantages enjoyed by hackers is that their crimes seem to noncomputer-savvy victims to be so exotic as to be unstoppable. This perception, even among law enforcement officers, means that oftentimes, those charged with protecting against cybercrime feel they should cede their duties to the "experts." Tosh.

As with the discussion of stricter penalties for cybercrime, it's important to remember that cybercrime is the act of appropriating, without permission, the property of another. The only difference between a cybercrime and a smash-and-grab theft is the type of window the criminal is breaking and the item of value he or she is taking. Crime is against the law. The problem we have currently is that officers, even those who specialize in cybercrime, have a difficult time articulating the elements of the offense, which means they have a hard time getting the prosecutor to file charges because the prosecutor is not confident he or she can get a judge to sign warrants because the judge is concerned that he or she will look like an idiot when he or she can't explain to the jury what just happened.

Those are radically different problems from "We need tougher sentencing guidelines for cybercriminals." Those are problems that can only be solved with training, time, and experience at every link of the chain, from the judge, to the prosecutor, to the investigators.

ATTACKERS' FAMILIARITY WITH HUMAN PSYCHOLOGY

In countless cases of cybercrime we have seen, the attackers' familiarity with the victims' subjects of interest has been a primary avenue for launching or better establishing a digital foothold from which to launch a convincing social engineering attack. In addition to the specific thing of interest that might get someone to open an e-mail—a note, for example, from someone you met at a recent industry confab, or a document sent by a colleague from another office—there's also a specialization on the part of professional cybercriminals in understanding human vulnerabilities.

We All Want to Help

There have been many studies[5] and books[11] showing just how far humans are willing to go to be helpful. Many social engineers aver that women are harder to scam than men, but in general, a good story and a charming smile will get you closer the vault than any collection of ninja suits and fancy rappelling gear.

A recent physical penetration test at a Fortune 500 firm in the midst of a cyber attack saw an 80 percent total success rate—that is, in 8 out of 10 attempts, researchers were able to concoct a story, deliver it, get into the building, get onto the network, and gain access to secret documents merely by telling a nice story and looking presentable. And this was a company at a moderately high state of alert. The "social engineering" aspect of the equation is often diminished in importance, but it serves to set the table stakes lower for the hacker.

Cybercriminals will know this about your people, and the attacks they launch will certainly—*certainly*—prey on this human vulnerability. It's not just untrained, unaware users who fall victim to this; even highly trained people who work at information security firms fall victim to it, all the time. So why would attackers stop doing something that patently works so bloody well?

To illustrate, let's look at two highly public hacks.

The Recruitment Spreadsheet Gambit: RSA Security

Cyber attackers will use guile and charm to get people to click on links or documents, even those that security programs have identified as "bad." Just

---

[5] A nice one, giving you a reason to wear a cast on your arm the next time you interview for a job, is Weinberg, N., "Social stereotyping of the physically handicapped." *Rehabilitation Psychology* 23(4), 1976, pp. 115–124.

ask information security vendor RSA. One of the most damaging breaches in history occurred when an employee of this large security company opened a document that came via e-mail (with the subject line "2011 Recruitment plan. xls") that security software had (correctly) placed into a spam folder. The document created a customized piece of malicious code that ultimately led to the total compromise of RSA's single most important annuity product, its SecurID two-factor authentication algorithm.[12]

The Idiot in the Window Affair: HBGary Federal

Early in 2011, Aaron Barr, the CEO of a security firm that created and sold custom security (read: malicious) software to government agencies, stated publicly that he had tracked down the names of members of an underground hacking group and would reveal the identities of its leaders soon. To oversimplify, some days later the hackers, pretending to be HBGary CEO Greg Hoglund, social-engineered the password to the firm's e-mail server by asking for it.[13] The resultant scandal saw Barr's disgrace and dismissal, and posed real challenges to HBGary as a company.

Until people like these stop falling for stuff like that, cybercriminals will continue to use these tactics to abet their crimes.

MOTIVATIONS AND EVENT-DRIVEN TRENDS

If you believe the security and even the mainstream media, the year 2011 saw a kind of "back to the future" when it came to cyber attacks. For almost a decade, cybercriminals were exhibiting all the signs in the world of mounting attacks which were financially motivated, but in 2011 the rise of the Anonymous and Occupy movements saw widespread attacks based, the attackers claimed, on philosophical, theological, political, and even humanitarian goals. This was the Renaissance of Hacktivism.

We believe that many of those participating in the attacks of 2011—whose targets ranged from large banks and manufacturing companies to law enforcement and intelligence Web sites—believe they were conducting acts of good in support of a stated agenda. We do not believe that those ultimately responsible for mounting the attacks had purely eleemosynary motivations. Instead, we believe that many of these attacks were financed by criminal groups, such as organized crime gangs, organized retail theft gangs, drug cartels, outlaw motorcycle gangs, and other groups that were intent on gaining intelligence and causing disruption in the apparatus sworn to fight them.

However, we don't dispute that nonfinancial motivations are claimed in many kinds of cybercriminal attacks, defacements, and breaches.

Politically Motivated Attacks

Attackers in support of a political cause or group use cyber attack methods to assist those they support. For instance, numerous cyber attacks were waged during and in support of recent uprisings in the Middle East. Examples include theft of law enforcement information in the United States, Columbia, the UK and other countries; attacks against the governments of Egpyt and Syria in support of anti-government forces and, most recently in the US, claims of hacks that would have released the tax returns of presidential candidate Mitt Romney, for the purpose of embarrassing his campaign. Examples of nation-state funded politically motivated attacks include data exfiltration and network penetration by Chinese hackers of the Office of the Dalai Lama.

Philosophically Motivated Attacks

Activists justify their behavior based on a perceived sense that the victim is complicit in crimes against some group, or acts in favor of some group, or benefits unfairly. Hackers attack the victim claiming to right wrongs. Examples include recent hacking attacks on PayPal for its refusal to support WikiLeaks by processing donations to it, and attacks against law enforcement agencies[14] in protest of alleged corruption or police brutality.

Financially Motivated Attacks

Financial motivation is the easiest to understand, and we spend most of our time in this book on this category of cyber crime.

SUMMARY

Cyber crime's face and composition has changed drastically in the past decade, and legislation has not kept pace with advances in cyber crime technologies. There is a wide range of motivators for cyber criminal activity, and the "cyber crime" bucket is not as neatly defined as some politicians might wish.

References

[1] A good place to start a search like this is the cybercrime practice at Cipher Law Group, at https://www.cipherlawgroup.com/index.php/en/legislation-update.

[2] http://www.whitehouse.gov/administration/eop/nsc/transnational-crime/strengthen-interdiction.

[3] www.dfinews.com/news/senators-call-change-cybercrime-law.

[4] www.govinfosecurity.com/articles.php?art_id=4033.

[5] Natarajan, M. (2010) International Crime and Justice. Cambridge University Press, pp. 158–160.

[6] Brehmer, B. (2005) "The Dynamic OODA Loop: Amalgamating Boyd's OODA Loop and the Cybernetic Approach to Command and Control". 10th International Command and Control Research and Technology Symposium: The Future of C2. McClain, VA. Available at www.dodccrp.org/events/10th_ICCRTS/CD/papers/365.pdf.

[7] See, for example, http://www.infosectoday.com/Norwich/GI532/Social_Engineering.htm.

[8] Selby, Spijk. (2011, spoken word) "My Dad Is a Paranoid Doodyhead", minute three: "You're always so paranoid and you never let me do anything! All the kids in school get to play multiplayer online games, and so what if my machine gets hacked, I don't have anything on it worth stealing anyway!".

[9] BBC News. (2004) "Passwords revealed by sweet deal." BBC News, April 20, 2004. Available at http://news.bbc.co.uk/2/hi/technology/3639679.stm.

[10] See http://news.bbc.co.uk/2/hi/technology/3639679.stm, and be prepared to laugh, and then weep.

[11] See, for example, Hadnagy, C. (2010) Social Engineering: The Art of Human Hacking. Indianapolis: Wiley.

[12] See http://boingboing.net/2011/08/26/security-researchers-trace-rsa-hack-and-secureid-breach-to-lame-excel-spreadsheet-phishing.html for the story, and our take on the brouhaha that followed at http://policeledintelligence.com/2011/06/08/security-is-not-about-marketing-until-it-fails/.

[13] See http://arstechnica.com/tech-policy/news/2011/02/anonymous-speaks-the-inside-story-of-the-hbgary-hack.ars/3 for a blow-by-blow; yes, we know we quoted Hoglund a few pages ago, which is consistent with the concept that (a) everyone gets hacked, and (b) getting hacked doesn't necessarily reduce the credibility your ideas might have in the security community [since (a) is true]. Unless you're STRATFOR.

[14] See, for example, http://policeledintelligence.com/2011/09/02/texas-law-enforcement-it-hit-by-criminal-attack-data-breach/ and http://policeledintelligence.com/2011/08/01/analysis-70-law-enforcement-sites-attacked/.

CHAPTER 2

Seasons of Change

INFORMATION IN THIS CHAPTER:

- From Experiment to Exposé to Exploit
- Timeline: Innovations, Intrigue, and Intrusions
- Propaganda and Lulz

CONTENTS

INTRODUCTION

No book that explores the emergence of illicit markets related to cybercriminal activity would be complete without examining the history of phreaking, hacking, and cracking. At its core, this historical record paints a picture that is heavily influenced by many factors. One of these is the driving force to know how something works and how to exploit it for use beyond the original intent of its designers. On its own this is not a foreign idea, yet it still warrants reference. *Why* we do what we do is as important as *how* we do it. In this chapter, we'll explore those ideas while demonstrating how they relate to the evolution of a new era of advanced criminal activity. Although this chapter won't provide an exhaustive account of the historical relevance of phreaking, hacking, and cracking, it will cover the key events that influenced what is occurring globally today, as well as provide insight into the hearts and minds that made these events a reality.

Some people love power. Others love prestige; seeing their name in lights on marquees, in magazine, and newspaper bylines, and in book reviews. Still others love money, and some love the ability to execute a plan, crafted after weeks, months, or even years of preparation, drawing less attention to themselves or their actions than the changing of a traffic light does. Some people—and for many this represents a real internal conflict often driven by cognitive dissonance when called to provide analytic insight—want nothing more than to watch the world burn, trading idealism and fanaticism for an otherwise sane means to an end.

In this chapter, we'll begin to explore these ideas that both vex and invigorate criminologists, psychologists, behavioral analysts, and profilers alike, while we try to reconcile them to the threats posed by the products and services crafted and delivered by cybercriminals. Often the root cause analysis ends at the beginning (or perceived beginning) of the evidence chain as it relates to systems and environments under our control. This book's goal is to provide greater insight into the prologue of the narrative materializing before us. Will we always successfully establish the *why* and the *who*? No, we will not. Yet, as long as we are dutiful and alert, we should, when presented with untainted data and evidence, begin to see patterns we can compare to data and evidence seen in other examples in the hopes of establishing points of confluence leading to the answers we seek.

We believe that *profit* and *gain* are the main factors that drive people to do what they do—right or wrong. But was this always the case? That is the question. Just as many factors lead a person to engage in cybercrime as in any other form of crime. The reasons are as diverse as the types of people participating in the activity and their motives. Variables we will explore in this chapter include, but are not limited to, the following:

■ Socio economic.
■ Cultural attitudes/beliefs regarding acts of vandalism or criminality.
■ Social/subcultural.
■ Psychological.
■ Economic.
■ Propagandist.
■ Philosophical.
■ Political.
■ Professional, subnational nonorganized/nonsyndicate.
■ Professional, subnational organized/syndicate.
■ Professional, subnational espionage (industrial).
■ Professional, subnational espionage (state).
■ Professional, nation state sponsored.
■ Revenge.

FROM EXPERIMENT TO EXPOSÉ TO EXPLOIT

Experimentation, fueled by natural curiosity, has played a vital role in the advancement of humanity, and particularly in the development of the sciences and the arts. Without a natural inclination toward trial and error, we would still be struggling with the impact of now easily addressed infectious diseases or how best to contend with the elements in order to ensure survival. No progressive act occurs without experimentation.

However, not all progressive acts have positive outcomes regardless of the role that experimentation plays, and cybercriminal activity has no exception. Because of this, it is important to recognize the role that agendas play in experimentation. Understanding agendas lets you access ideas and variables that influence motive. Gaining insight into the "cause" of an event or action lets you further explore the agendas and the intentions of those behind them, often yielding a wealth of data. Many times it is the small things that lead to revelation. Similarly, it is often the seemingly small or insignificant things that go unnoticed and thus unattended, allowing cybercriminals to continue exploiting them. These proverbial crumbs, when left ignored, encourage the proliferation of vermin resulting in profound infestation.

TIMELINE: INNOVATIONS, INTRIGUE, AND INTRUSIONS

People have tried to develop timelines that accurately depict the history of phreaking, hacking, and cracking several times through several different forms of media. After careful consideration, we felt it was best to approach this topic at a relatively high level, as there is much debate regarding many of the "noteworthy" events that have occurred and the circumstances surrounding them. We have sought to establish accuracy and authenticity regarding these events in an effort to demonstrate the historical ties between experimentation of evolving systems and technologies and their exploitation on a global basis.

The First among Equals

Our timeline begins with the events surrounding the life of John Nevil Maskelyne. Maskelyne was born on December 22, 1839, in Cheltenham, Gloucestershire, England. A watchmaker by trade, Maskelyne had a curiosity and inclination toward experimentation that caused him to publicly denounce a performance given by the famed American illusionist team known as the Davenport Brothers after he saw how the Davenports' spirit cabinet illusion worked. With the help of his friend George Alfred Cooke, a cabinetmaker, Maskelyne created a cabinet that replicated the Davenports' illusion, exposing it to the public in Cheltenham in June 1865.[1] Maskelyne and Cooke noted the crowd's approval of their work along with the acclaim it earned them, and decided to develop their *own* act.

The pair became stage magicians, developing and performing illusions, among them the ubiquitous levitation. In 1894, Maskelyne wrote a book titled *Sharps and Flats: A Complete Revelation of the Secrets of Cheating at Games of Chance and Skill*; this renowned book is still considered a classic exposé of illicit gambling methods. Maskelyne's later associations included membership in the famed

Magic Circle of London, founded in 1905 and considered by many as the world's leading institute for the study of magic. In 1914, Maskelyne founded the Occult Committee, whose charter was to "investigate claims to supernatural power and expose fraud". Though Maskelyne's work focused mainly on the exposé of illusion, stagecraft, and fraud, we believe it is not only relevant but also paramount to the topic of this chapter.

The Birth of the Hack

In addition to the work noted here, Maskelyne also was the father of another famed British stage magician, Nevil Maskelyne. Born in 1863, Nevil grew up in a world surrounded by intrigue, experimentation, and exposé. Learning his tradecraft from his father and his father's colleagues, Nevil would eventually continue his father's work at the Egyptian Hall in Piccadilly, London, a museum commissioned by the William Bullock collection of antiquities and curiosities and later becoming synonymous with spiritualism and magic. Among Nevil's other areas of interest was then the bourgeoning technology of the wireless telegraph. While working within this discipline, Nevil came into his own as history's first "hacker."

In June 1903, a curious quiet fell on the audience in the Royal Institution's celebrated lecture theatre in London. At the front of the crowd stood the noted British physicist John Ambrose Fleming, making final adjustments to strange-looking instruments in preparation for what would be a breakthrough in modern communications technology. (This demonstration would set the tone for technological advancement for the next 100 years, and Fleming would go down as one of its forefathers.) Fleming examined the equipment and prepared to demonstrate this new technological breakthrough: a long-range wireless communication system developed by his employer, the celebrated Italian pioneer of radio technology, Guglielmo Marconi. The demonstration's purpose and goals were clear: Demonstrate and showcase publicly, for the first time, that Morse code messages could be sent and received wirelessly over long distances, a concept that made many people both suspicious and hopeful.

While Fleming made his final adjustments to the instruments trusted to his care, Marconi was preparing to send a signal from a telegraph station located atop a cliff in Poldhu, Cornwall, England. Marconi was no stranger to controversy or skepticism; his advanced work, alongside that of his mentor, Nicola Tesla, is credited for advancement in both wireless communications and energy transmissions leading to a great deal of speculation and conspiracy theories. [2] On this day, he was ready to demonstrate what he believed was a secure and private mode of communication. Before the team of Fleming and Marconi could initiate their demonstration, however, the machine in the lecture hall began to receive and present a message. Initially, one word was received and repeated over and over again. The word was *rats*. [3] As quickly as that transmission had been received, it changed and then began to produce a limerick crafted at the expense of Marconi himself! In the poem the following was transmitted: "There was a young fellow of Italy, who diddled the public quite prettily."

At the time, the transmission was, as one might have expected in 1903, a mystery to everyone who witnessed it. Fleming could not account for it. Neither could Marconi. It was clear that Marconi's assertion regarding the privacy and security of this technology, though well meaning, was not entirely accurate. Fleming and Marconi had been hacked. A precedent had been set that would set in motion events on a global basis that would evolve from prank to profit. Experimentation would give way to exploitation, and no communication system, or any system connected to said communication system, would ever truly be private or secure. But why the hack experiment being conducted by Fleming and Marconi? To what end?

In modern parlance you might think the hacker was in it for the "lulz," but was that all there was to this compromise? [4] Over the years, numerous people have speculated about the events that led to the now infamous hack. The general consensus is that the events of that fateful day in June 1903 had their roots in work conducted by Heinrich Hertz. In 1887, Hertz achieved a significant goal in proving the existence of the electromagnetic waves described by British scientist James Clerk Maxwell. Maxwell had developed a scientific theory for the express purpose of explaining electromagnetic waves. Through his work, Maxwell had observed that electrical fields and magnetic fields often combine to form electromagnetic waves. Maxwell's theory asserted that individually, an electrical field would not move by itself. Similarly, he observed that magnetic fields likewise lacked the capability to propel themselves without aid. Maxwell asserted that through manipulation of magnetic fields one could, by proxy, achieve change in electrical fields and vice versa.[5] In his experiment, Hertz proceeded to discharge a capacitor into two separated electrodes and then ionized the air gap between them, leading to the creation of a spark. A subsequent spark materialized between two electrodes a few meters away, thus proving Maxwell's theory and giving birth to "Hertzian waves." It was later theorized that these bursts of energy could be leveraged for the transmission of Morse code.

Enter Guglielmo Marconi and his company. Marconi was at the forefront of the wireless communications industry. He claimed that through his solution and instruments he could arrive at a configuration that no other instrument or system could tap into. The implication was that his solution, provided there was not another within a reasonable proximity that shared similar configuration and instrumentation, was "secure" and "private." The diatribe ended as swiftly as it began, however; mere moments prior to Marconi's communications arriving.

Though Fleming and Marconi continued their demonstration, the impact and damage of the previous phantom attack was done. It was clear that Marconi's assertion regarding his solution's security and privacy was overblown. The implications were grave as it became obvious that this could lead to breaches and result in information meant for specific parties falling into the hands of those for whom it was unintended. Marconi, to his credit, didn't acknowledge the incident, preferring to let history be its judge, but Fleming did. He sent letters to newspapers in London, and one letter in particular that found its way to the offices of *The Times of London* saw Fleming appeals to the readers of the paper to help him and Marconi locate the individual responsible for the intrusion. Four days after the incident occurred, a letter materialized and was printed in *The Times of London*. In the letter, the author justified his actions, citing that through them he had successfully identified security holes within Marconi's system, thus serving the greater public good. Unlike in the latter portion of the 20th and early 21st centuries, where parties responsible for similar actions hide behind pseudonyms and "handles," the author identified himself proudly.

He was 39-year-old Nevil Maskelyne. Like his father, Nevil was trained in the magic arts. However, his passion and interests lay more in wireless technology. In fact, Nevil was so interested in the technology that he taught himself its principles, many of which, no doubt, were defined or refined by Marconi and his mentor, Tesla. Nevil's personal experience in Morse code, incorporated in his stage act, was well known, as was his incorporation of spark-gap transmitters for the ignition of explosives onstage.

Nevil Maskelyne continued to work toward pioneering wireless communication transmissions, allegedly sending messages in 1900 between a ground station and a balloon 10 miles away. However, like many of Marconi's contemporaries, Nevil became frustrated by the broad patent grants Marconi had received, leaving him less than enthralled with Marconi.[6] The Eastern Telegraph Company had recruited Nevil to conduct industrial espionage on Marconi after his successful transatlantic wireless transmission on December 12, 1901.[7–9] In 1902, Nevil revealed in an article penned for a trade journal that he had identified security weaknesses within Marconi's solution and had in fact intercepted messages. How did Nevil achieve this when Marconi had what was believed to be a patented, secure solution for wireless transmission?

The answer lies in how Marconi described his solution. He had asserted within his patent claim that his solution enabled tuning a wireless transmitter to broadcast on a precise wavelength, thus implying confidentiality of transmission via confidential or secured channels. Unfortunately, science was not on Marconi's side, and Nevil understood this. Using an untuned broadband receiver, Nevil was able to listen in on the transmission. His desire to undermine Marconi led to his interference in Fleming and Marconi's transmission. When the opportunity presented itself Nevil was prepared and executed his intrusion in much the same way that hackers do today.

Hacking and Eavesdropping in Times of War

The decades that followed Nevil Maskelyne's hack of the Fleming/Marconi demonstration in London saw great change in Great Britain, Europe, and beyond. The world had experienced the Great War and saw for the first time the true reach and extent that superpowers had as a result of industrialization. The map of Europe had once more been crossed and crisscrossed by armies rallying to the aid of those with whom their leaders were aligned and called upon to defend, and in like fashion to the aid of those to whom they had been bound through alliance for the purpose of defeating a common enemy. Advancements in technology were realized on the battlefield and off in ways previously thought impossible. Concepts such as "war of attrition"[10] (warfare focused on human lives and their subsequent exhaustion as a vehicle toward victory, trench warfare, and chemical warfare, in addition to innovations in air, sea, and land warfare) saw introduction and adoption on a massive scale.

Fifteen million soldiers lost their lives[11] and entire nations reeled at what would become known as The War to End All Wars.[12] Advancements in communications and intelligence were also realized during this period. Though rising to greater prominence and ubiquity rapidly during World War II, communications intelligence began to play a more important and prevalent role in the war effort against the Central Powers. Though radio technology was still largely in its infancy, it quickly became extremely important strategically. Nations vied to obtain and retain control of undersea communications cable traffic. Upon declaring war against the Central Powers, one of Great Britain's first acts was to disable the undersea cables previously owned and operated by the Germans. In doing so, the British forced the hand of the Germans toward the use of the radio and seized on the opportunity to intercept communications as they saw fit in a manner reminiscent of Nevil Maskelyne (work that would eventually become the domain of the cryptanalysts working in room 40, or NID25, within the British Admiralty). Since that time, the disablement of secure wired communications has become paramount in ensuring superior intelligence during times of conflict and war. At this time, much work was being done in addressing the challenges associated with keeping communications secure. Inventors in many countries around the world began to reach a similar conclusion: If they employed a completely randomized key sequence that contained no repetitive patterns they could effectively achieve an unbreakable, polyalphabetic substitution cipher (the polyalphabetic substitution cipher is based on a substitution that utilizes multiple substitution alphabets like the Vigenère cipher). These advancements gave way to rotor cipher devices that were later eclipsed by the most advanced cryptographic device of its time: the Enigma machine.

Hacking for the Greater Good

The Enigma machine needs little introduction in most information security circles, as its relevance and importance to both information security and world history is well known. What makes the Enigma machine relevant to this chapter is that it was compromised and subsequently exploited in a manner that was previously considered impossible. Developed and crafted by the German engineer Arthur Scherbius near the end of World War I, the machine offered an opportunity to demonstrate the importance of cryptography to both military and civilian intelligence applications.

Scherbius was born in Frankfurt am Main in 1878. Trained and educated as an electrical engineer, Scherbius held many patents, one of which was received in 1918 for the device he named Enigma, after the Greek word for "riddle." Several other electrical engineers and inventors were exploring the principles and concepts of rotor cipher devices at the same time that Scherbius pursued his design. Though many of these inventors made significant progress with their designs (for example, Hugo Alexander filed for and received a patent for his rotor cipher device in the Netherlands in 1919[13]), all of them failed to identify and capitalize on an addressable market for their designs. Scherbius also had difficulty initially gaining a competitive foothold for his technological design, until the German military inquired about his work and its potential application in ensuring the "privacy" and "security" of communications.

Production of Scherbius's machines began in 1925 and the machines were deployed starting in 1926. This period of German history was significant, as it saw a Germany struggling to regain a sense of balance and identity after suffering what it perceived as a humiliating defeat in World War I. Known as the era of the Weimar Republic, it was during this time that Germany was being plagued with political extremists on both the left and the right, hyperinflation, and pressure from the victors of World War I who had attempted to restructure Germany's payments to their advantage twice through the Dawes Plan and the Young Plan, respectively. This instability would ultimately lead to an election in 1933 that allowed for a political extremist group and its leader, Adolf Hitler, to ascend to power via election, and subsequently usher in a new government.

Prior to the events of 1933, the French and Americans also struggled with intercepting and deciphering encrypted German radio transmissions. The failure to decipher this message traffic left the victors of World War I in an awkward position. Fully aware of the conditions of the Weimar Republic, its economy, and its divided populace, the Allied Forces knew only too well what a Germany that was once more economically strong and unified in mind, cause, and spirit was capable of. As a result, it was in the best interests of the Allied Forces, Europe, and the world to (for a time) *hope* that their former foes would remain downtrodden. The matter remained complicated by the continuing lack of insight into sensitive German communications now securely encrypted using the Enigma machine.

In 1932, the Polish Cipher Bureau[14] initiated a project to break the ciphers present within Scherbius's Enigma machines that were now in use throughout the German military.[15] Due to the lingering fear of a refocused, revitalized Germany bent on invasion the Polish Cipher Bureau spent the next seven years working to overcome the Enigma ciphers. Though there appears to be some debate with respect to whether the Polish Cipher Bureau had assistance in its project from the French Intelligence Service,[16] its presentation of its findings to members of both the British and the French Intelligence communities took place a mere five weeks before the outbreak of World War II, on July 25, 1939, in Warsaw, Poland.

The work of the Polish Cipher Bureau would aid the Allied Forces immeasurably in their struggle against the Axis powers of Germany, Italy, and Japan throughout World War II. The hack of the Enigma machine would set a precedent for hacks conducted on behalf of the greater good. However, as we will discuss in later sections of this chapter, many times what is considered to be the "greater good" is highly debatable.

Exploration of Systems: Phreaks, Geeks, and Cereal
While the hack of the Enigma machine is a good example of an early form of hacking, many describe the situation involving AT&T and its interoffice trunking systems as the birth of modern hacking.

AT&T had a problem, one that resonated at precisely 2600 Hz. On two separate occasions, AT&T published in the *Bell System Technical Journal* articles that described its processes for its interoffice trunking systems, signaling, and frequencies.[17][18] The first piece, described the actual processes, signaling, and frequencies. The second piece, titled "Signaling Systems for the Control of Telephone Switching," provided the proverbial keys to the kingdom. In this article, the authors revealed the frequencies used for the digits employed within the Bell interoffice trunking systems as actual routing codes.

Armed with the data contained within the first article anyone with an elementary knowledge of electronics could take control of the public switched telephone network (PSTN). AT&T had inadvertently described *how to hack* its network, thus ushering in the era of the *phreaker*, the generic term used for those who, in an unauthorized manner, exploit the PSTN in ways unintended by its architects and owners. As soon as it became apparent that this information had been made publicly available AT&T began to take precautionary measures, but it was too late. The information was out in the public, and there was now no simple way to get it back or, worse, disavow knowledge of it being in the public in the first place. Soon after this occurred a whole generation of phreakers came into their own; stories began to materialize of phreakers discovering the now famed "2600 Hz tone" and how to teach others to mimic it through natural means. Some, such as the infamous Josef Engressia (a.k.a. Joybubbles), who was born blind and with perfect pitch,[19] were able to mimic the tone perfectly without the aid of any electronics. Others, such as John Draper, utilized unnatural means to emulate the tone. Draper eventually rose to a position of notoriety and became an icon in phreaker and hacker communities alike under the moniker of Captain Crunch due to his use and adoption of the now infamous plastic whistle that appeared in boxes of Cap'n Crunch breakfast cereal during the 1960s.

AT&T leveraged the tone as a consistent signal used in classifying long-distance telephone lines or "trunk lines"; once a person mastered the tone, he or she could reset the lines in order to use them. Later techniques, such as trunk flashing, would emerge as well, allowing phreakers to manipulate the interoffice system to route calls. These techniques, in addition to other similar tone-driven techniques, led to the development of electronic devices that could generate the tones produced and employed by a Bell Systems telephone operator's dialing console to switch long-distance calls. These electronic devices were known as Blue Boxes. Other color-coordinated "box" technology soon emerged in the form of Red Boxes, which allowed the user to generate tones that simulated the sound of coins being inserted into a public pay telephone in order to fool the system and complete a free telephone call, and Black Boxes that allowed phreakers to outfit their telephone to complete a free-of-charge call to any other telephone that was similarly equipped with a small, third-party circuit. Draper became a master Blue Boxer along with other noteworthy phreakers, including Apple cofounders Steve Wozniak and Steve Jobs.

Shortly thereafter, an article appeared in a 1971 issue of *Esquire* magazine written by Ron Rosenbaum and titled "Secrets of the Little Blue Box."[20] The resultant surge of interest in phreaking was staggering. Suddenly, as though without warning, an entire subculture emerged around the Blue Box that aided in perpetuating the fame and notoriety of Draper and the Legion of Doom[21] (discussed shortly). In this way, phreaking became the direct ancestor of modern hacking and cracking.

Although some people debate the semantics of the terms *phreaking*, *hacking*, and *cracking*, the base quality shared by all of them can be surmised in the following manner: unauthorized, unlawful access to or exploitation of systems outside of the original design intents. Seemingly innocent, the activities associated with phreakers and phreaking were anything but innocent. Many times they were related to ill-conceived practical jokes, yet all of them constituted theft of service and would pave the way for generations of hacking enthusiasts around the world inclined to experiment, at times to their peril.

Organization Skillz: The Birth of Hacker Culture

Phreaking inspired many would-be exploiters' curiosity, and as we mentioned in the preceding section, some went on to great heights forming large, revolutionary technology corporations that continue to leave their imprints on our lives, while others went on to be immortalized in the mythology of a subculture fueled many times by hyperbole. Yet in the mid-1980s something began to change. Hackers, phreakers, coders, and crackers began to meet and, to the best of their abilities, they united for the express purpose of proliferating their art and craft many times at the expense of others. Though the conventional wisdom of the day did not view it so, these early hacker meetings would lead to the genesis of subcultures within a growing, morphing subculture. At the precipice of this evolution stood the early phreaking/hacking groups whose ability to attract and assemble likeminded individuals rivaled and excelled beyond their ill-fated predecessors. In 1984, two such groups sprang into existence: Legion of Doom and Masters of Deception. Both left their imprint on their respective subcultures and the future information security industry in whose lineage they would be included.

LOD and MOD

One of the most notorious of the early phreaking/hacking groups, Legion of Doom (LOD) was founded by an anonymous individual using the alias or handle of "Lex Luthor."[22] Though we now know this person's real name— Vincent Louis Gelormine—for many years both he and his identity remained a carefully guarded secret within the community in which LOD was a key participant. The group began to make a name for itself quite quickly, shooting to notoriety within the underground through its hacking escapades and prolific publishing of the *Legion of Doom Technical Journals*. The journals contained guiding principles, code, and programming examples as well as other information deemed key to nascent hackers and crackers around the world.

Soon LOD garnered the ire of a rival hacker group, Masters of Deception (MOD). MOD was formed in 1990–1991 by a hacker using the alias Acid Phreak (real name: Eli Ladopoulos). Ladopoulos had begun to recruit members initially during meetings on loop-around test lines. Those exchanges led him to connect with other young hackers going by the aliases of Scorpion (real name: Paul Stira) and HAC, to form what can only be considered as an early, legendary collaboration among young hackers who sought to compromise and exploit systems such as the Regional Bell Operating Company (RBOC) switches and control systems (minicomputers and mainframes) used in network administration.

The group had adopted the acronym "MOD" in mockery of LOD whom many members of MOD believed to be "lost" with respect to the ways of the underground hacker. Some time during 1990 and 1991, the tensions between the two groups escalated into what is commonly referenced as the first real "hacker war." This war would see these young hackers take extreme measures against one another using the Internet, X.25, and telephone networks to deliver their blows. The animosity rose to heights so great that members of LOD were said to have even launched an Internet security consultancy to assist corporations victimized by MOD. The hacker war also saw some members of LOD, such as Phiber Optik (real name: Mark Abene), shift their allegiances to MOD. In the end, indictments were handed down to members of LOD beginning in 1990 (perhaps most notable was the indictment of LOD member Leonard Rose, who was convicted in 1991 of illicit use of Unix 3.2 source code owned by AT&T, which Rose allegedly stole from AT&T using Trojan horse programs from May 1988 to January 1990),[23][24] and MOD in 1992. While LOD and MOD fought it out on the highways and byways of the primitive Internet and PSTN, another group had also come online and into its own. With a bold, clear message and a willingness to see their causes succeed at all costs, the Cult of Dead Cows (cDc) was born.

h@X0r b0v1n3

Writing a section in a book on cybercriminal activity that discusses the infamous hacker group Cult of Dead Cows (cDc) is difficult, to say the least. The group's history, actions, tools, and causes are literally enough to fill an entire book dedicated to it and its legacy. The purpose of this section, like the others, is not to glorify or deify this hacker group or any of the others mentioned previously, but rather to articulate the maturity and strength of numbers that has emerged in this space and, on one or more occasions, has been put to use for purposes falling short of the greater good.

Founded in Lubbock, Texas, in 1984 by three aspiring hackers and computer aficionados, Swamp Ratte (a.k.a. Grandmaster Ratte), Franken Gibe, and Sid Vicious, cDc gathered momentum and clout for the rest of the 1980s and into the 1990s through a number of different activities, most notably in the development of tools useful in enumerating networks[25][26] and hosts and for their role in hacktivist activities.[1] Over the years, the group expanded its interests and tastes to include teams devoted to the perpetuation of its technical, social, and political agendas. cDc also gave birth to one of the first underground "elite" organizations known as the Ninja Strike Force (NSF),[27] which has, since its inception, prided itself on its secrecy and elitism. Where other hacker collectives failed to embrace one another and cross-pollinate one another's ranks, cDc flourished. Notable members of the now defunct L0pht hacker collective, including DilDOG (real name: Christien Rioux) and Mudge (real name: Peiter Zatko), established themselves with the cDc's now infamous ranks alongside IOERROR (real name: Jacob Applebaum),[2] who has ties to WikiLeaks founder Julian Assange and former U.S. Army Intelligence Community member PFC. Bradley Manning,[28] who has been arraigned and is awaiting trial at the time of this writing.[29] Though no stranger to controversy, cDc and most of its members have remained relatively unscathed and unindicted throughout the group's lifespan. Through their ties to nouveau hip hacker collective and activist organizations, Anonymous and LulzSec have recently drawn attention to themselves based on the group's previous involvement in hactivist activity.[3, 4]

4chan and Anonymous

On October 1, 2003, an all image-based bulletin board Web site called 4chan (www.4chan.org) was launched.[30] 4chan had been inspired by similar sites in Asia, most notably the colossal Internet forum 2channel (www.2ch.net). The site has more than 600 boards addressing a diverse range of subjects. The site's creator, Christopher Poole, also known as moot, had ties to Raspberry Heaven, an Internet Relay Chat (IRC) community that was originally composed of members of the Anime Death Tentacle Rape Whorehouse subforum known as Somethingawful.com. Poole's site allows contributors to post content anonymously. The site is largely text-based and driven heavily by Internet memes. If a site visitor does not complete the Name field when posting content or responses to other users, the content is automatically credited as "anonymous." Within the greater 4chan community, the /b/ channel was and remains one of its most popular user forums. It is within the /b/ channel that perhaps the world's most infamous hacker collective, Anonymous, took root and blossomed.

Anonymous hosts an informal culture that complicates efforts to define the culture and leadership of the organization. It possesses no formal hierarchy that aids its members in achieving plausible deniability while also encouraging its faux altruistic charter. Anonymous fully embraces and has, in certain respects, redefined the term *hacktivist*, first defined by cDc member Omega in 1998.[31]

The portmanteau *hacktivist* combines *hacking* and *activist* and its invocation implies (especially among the 4chan/Anonymous community) that they are hacking and cracking for a cause as opposed to absent of one. This is one of the more interesting and potentially dangerous aspects of this organization. Causes can be geopolitical (a protest against the G20 or G8 Summit, for example), theological, and/or philosophical, as we'll discuss later in this chapter. Actions (regardless of the lawlessness involved) can and often will be justified in due course to ensure cohesive alignment with the "group think" that permeates the organization. Those who are active and willing participants in these actions (colloquially referred to as Ops) have certain common beliefs that guide their activity and the vigor with which it is delivered. Chief among their concerns is corporate greed and corruption, followed by political greed and corruption.

One of the strengths of Anonymous is its reluctance to recruit, like many other hacker collectives. Instead, Anonymous encourages anyone concerned with the causes it is concerned with to join the organization in the IRC channels that it leverages for its Ops. The organization, as we will discuss throughout this section, is well known and just as likely to partake in or orchestrate an in-person demonstration, as it is an online one.

Take Project Chanology, for example. In 2008, Project Chanology began in earnest after a video from the Church of Scientology was posted on YouTube without the Church's permission on January 14 of the same year. The video was a testimonial piece featuring a gregarious Tom Cruise discussing his views on Scientology, his beliefs as to why the world needs Scientologists, and the insights that Scientologists share among themselves that are seemingly missed by those who exist outside their belief system.[32]

The video made the rounds on the Internet and eventually became the fodder for mockery and jest on the 4chan /b/ forum and within various IRCs. Understandably, the Church of Scientology moved to have the video removed from YouTube, citing copyright infringement. The resultant response from the underground was the shot heard around the world for those espousing and embracing shared ideals regarding the matter. On January 21, 2008 a video was posted in response on YouTube titled "Message to Scientology." It was credited to Anonymous, and so began the project now commonly referred to as Project Chanology. Soon thereafter a press release emerged explaining the intentions of the group now known as Anonymous and their current project, Project Chanology.[33] Within the press release the organization asserted that the Church of Scientology was a dangerous organization that sought to exploit its members financially, not hesitating to engage in threats, blackmail, and an assortment of other abuses when it served their purposes. The press release went on to say that the attempt to have the video removed was a violation of the right to free speech. On January 28, 2008 Anonymous posted a Call to Action on YouTube that called for protests to be held outside Church of Scientology centers around the world on February 10, 2008. Reports of distributed denial of service (DDoS) attacks against the Church of Scientology Web site were noted. The physical protests called for in the Call to Action video began and continued throughout the year. The Anonymous protesters donned the Guy Fawkes masks popularized in the motion picture *V for Vendetta*. The masks served practical as well as symbolic purposes, as they enabled the protesters to ensure their identities were preserved from the prying eyes of the Church, which was known to take action against dissenters of Scientology, calling them "Suppressive Persons."[34][35]

In 2009 Anonymous once more sprang into action, this time in response to protests of the June election of Mahoud Ahmadinejad, president of Iran. Anonymous Iran was formed shortly thereafter. This marked the first public online project between Anonymous and another organization, The Pirate Bay, an illicit torrent search engine site based in Scandinavia. The Anonymous Iran project offered Iranians a forum and outlet to the world that was safely guarded during the crackdowns by the Iranian government on online news stories of the riots. Shortly thereafter, Anonymous launched Project Skynet to fight Internet censorship worldwide; this project remains in effect to this day.

In February 2010, the Australian government was working to pass legislation that would restrict certain types of adult content that the public may find offensive or may be indicative of the exploitation of minors. On February 10 of that year, Anonymous launched an operation designed specifically to demonstrate the group's displeasure at what members believed to be oppressive legislation. DDoS attacks were launched against numerous Australian government Web sites. In September of that year Anonymous launched Operation Payback, which it enacted in response to the Motion Picture Association of America's (MPAA) and the Recording Industry Association of America's (RIAA) alleged contracting of an Indian software firm to launch DDoS attacks against The Pirate Bay and other Web sites participating in illegal file sharing activity.[36] In retaliation, Anonymous launched its own DDoS attacks toward the MPAA, the RIAA, and Indian software firm AIPLEX.

The operation continued in December of that year; however, the targets had become global credit card corporations MasterCard and Visa, online payment corporation PayPal, the Bank of America, and Amazon.com. Anonymous enacted this series of assaults due to these corporations' decision to no longer process charitable donations to the WikiLeaks Web site. WikiLeaks had become a veritable clearinghouse of illegitimately gained information about allegedly corrupt government and corporate activities around the world. Registered on October 4, 2006, the site went live in December 2006 with its first documents. Represented predominantly by Julian Assange, the site continued to make its presence known through the work of volunteers and its founders, among them Assange, Chinese dissidents, journalists, mathematicians, and a vast array of technologists working for startup companies in the United States, Taiwan, Europe, Australia, and South Africa. By December 8, Web sites belonging to MasterCard and Visa were brought down through the efforts of Anonymous's DDoS attacks. On several occasions throughout the remainder of 2010, Anonymous would be wrongfully (or rightfully, depending on whether your beliefs and understanding of the organization allow you to dismiss the potential of "false flag" operations being conducted by the group to divert attention to or from it) of attacks resulting in a total site and database compromise of Gizmodo.com, Gawker.com, and Jezebel.com (it was later learned that another hacker collective, Gnosis, had its site on Gawker.com and took credit for that attack), most likely through network enumeration, vulnerability assessment, and a combination of brute forcing, dictionary, and rainbow attacks.

Operation Payback also saw Anonymous wield its capabilities against alleged corruption in the Zimbabwean government.[37] In this case, Anonymous threatened to paralyze Web sites belonging to parties that acted against WikiLeaks. The Zimbabwean government came under fire due to documents released through WikiLeaks that alleged that Grace Mugabe (wife of President Robert Mugabe) had made "tremendous profits" through trading in illicit diamonds. Anonymous took offense at President Mugabe's statements on taking action against WikiLeaks and anyone who posts documents to it. The reaction of the Anonymous collective was to launch DDoS attacks at Zimbabwean Web sites.

Several patterns begin to emerge when you analyze the actions of the Anonymous collective. Perhaps the most profound of these is that Anonymous, due to its loose confederation of individuals and groups bound by no specific allegiance or cause, aside from that which is at the core of the operation at hand, in many respects defies categorization. This inability to consistently characterize the behavioral patterns of Anonymous as a whole, independent of the individuals participating in one or more operations, makes the organization quite noteworthy and deserving of study.

Why would this be relevant in a book on illicit blackhat economies? For starters, by definition Anonymous refuses to operate in any way that would revoke individual or group anonymity. The potential for exploitation of this condition by seasoned criminals operating within organized crime syndicates, subnationally motivated entities working on behalf of criminals or nation states, or nation states themselves as part of tactical and strategic false flag operations is great. So great, in fact, that it warrants greater analysis on all fronts as the potential for vulnerability is present within the group's framework and is likely being exploited.

January 2011 saw Anonymous's activity rise to new heights, aiding in what has been called the "Jasmine Revolution" and the "Arab Spring."[38] Anonymous had launched two DDoS attacks in succession against the Tunisian Stock Exchange and Tunisian Ministry of Industry, respectively.[39] The reason for these attacks stemmed from a decision the Tunisian government made to try to restrict the Internet access of its citizens due to many having been critical of the government in active blogging and micro blogging initiatives. At the thought of a sovereign state whose government doesn't support freedom of speech, Anonymous enters and does so with a vengeance.

By the end of January, Egypt's regime had become the next target of opportunity for Anonymous's latest operation. The goal was clear: Aid the citizens of Egypt in their struggle to remove Egyptian President Hosni Mubarak from the position he held for more than three decades. The moment the Egyptian government decided to block access to Twitter, Anonymous struck by launching DDoS attacks against Egyptian government Web sites.

The year 2011 also saw Anonymous contend with information security professionals and firms publicly during an event that is perhaps one of the most widely debated and least understood of the past five years in the information security industry. Security researcher and HBGary Federal CEO Aaron Barr had an interest in social media networking. He had spent countless hours observing the nuances and potential for the technology's applications in a variety of ways. He, like many others, had also developed an interest in the hacker collective Anonymous and began to collect data and conduct research related to the group and its members. Barr's intent was academic, and as a sign of that bent toward academics, Barr had submitted an abstract to the 2011 Security BSides San Francisco Conference. A lot has been written about Barr's story and how the events unfolded. Some accounts are more debatable than others. Nevertheless, Barr's interaction with Anonymous is noteworthy as it demonstrated an intriguing insight into certain members of the collective. This became evident when the abstracts for the conference were published. Barr's abstract discussed his research in social media analysis and the potential applications in unraveling the identities of persons on the Internet. Anonymous's response was swift and calculated. The group proceeded to exploit vulnerabilities present within the HBGary Federal Web site (specifically a SQL injection vulnerability) gaining access to the site's database. Once the site was compromised, the hacker collective gained access to usernames, e-mail addresses, and password hashes. It is believed that Rainbow Tables were used to crack the MD5 hash algorithms that led to eventual access of the entire database.

HBGary Federal and Barr paid a significant price for the research and *feared disclosure* of the identity of members of the Anonymous leadership. Anonymous resorted to tactics and strategies that lacked the honor and integrity the group so often espoused by taking such personal action against Barr, HBGary Federal, and individuals associated with the compromise.

Not long after the events that surrounded HBGary Federal and Barr took place, Anonymous turned its attention to other targets, this time perhaps one of the world's largest online entertainment providers. Sony's PlayStation Network (PSN) had made a decision regarding one of its subscribers which would come back to haunt the company. Sony had decided to ban one of its PlayStation

Network customers for jailbreaking and modifying his PS3 console. In reality, George Hotz, also known as GeoHot,'[40] was much more than a simple user of the Sony PlayStation Network. Hotz, a gifted young man, had attended a prestigious magnet school[41] in New Jersey near his childhood home. Additionally, he had attended the Johns Hopkins Center for Talented Youth[42] and the Rochester Institute of Technology.[43] He gained notoriety in 2009 and 2010 for his work in jailbreaking Apple iPhones, and released a series of utilities for jailbreaking the Apple devices. The first utility, purplera1n,[44] allowed the user to jailbreak iOS Version 3.0 on iPhone and iPod Touch devices. Jailbreaking was achieved by editing the device's firmware while the device was in Device Firmware Upgrade (DFU) mode. Once the firmware was patched, the user could install utilities such as Cydia,[45] an alternative to Apple's Apple Store for jailbroken devices, or Rock App, a Cydia alternative. This process allowed users root access and the ability to add and remove applications at their discretion. Purplera1n was followed by limera1n[46] and subsequently by blackra1n. In late 2009, while he was busy working on his iPhone jailbreaking utilities, Hotz announced his intention to hack the Sony PlayStation 3 console (a system which many regarded as the only truly secure gaming system of its generation). On January 22, 2010, Hotz announced that he had successfully hacked the PS3 through enabling read and write access to the machine's system memory. He also stated that he was able to achieve access to the machine's hypervisor, something most people considered to be near impossible. Hotz went on to detail through his blog posts his progress and the features and functionality his research provided to users on the now jailbroken PS3. These functions included access to homebrewed apps and PlayStation 2 emulation.[47] On January 26, 2010, Hotz released his work to the public, noting that it required the presence of the OtherOS feature from PS3 models. Also required was a Linux module and access to the device's hypervisor via bus glitching. In spring 2010, Hotz posted a video of his progress on the Internet, showing a PlayStation 3 running with OtherOS feature-enabled on firmware 3.21 (which Hotz customized and dubbed "3.21OO"). He later posted the root keys of the PlayStation 3 on his Web site.[48]

Things became complex quite quickly for Hotz shortly thereafter.[49] Sony took legal action to have the keys removed from the Internet, and this eventually led to Sony filing suit against Hotz. The case was settled,[50] but not before Anonymous lashed out at Sony for its actions toward Hotz. Through launching DDoS attacks, Anonymous was able to successfully bring down the PlayStation Network and various other Sony Web sites. Sony would spend several weeks repairing the damage and would lose approximately $171 million.

Anonymous demonstrated an ability to impact and influence the actions of nation states and corporations, but it did not stop there. In July 2011, a rash of conversations began in the media regarding alleged corporate corruption on

Wall Street.[51] A movement was being birthed with a call for mass protests beginning July 17th, 2011. On August 23, 2011, Anonymous formally expressed its support of the movement, known as Occupy Wall Street, by posting another video on YouTube. Anonymous has remained close to the Occupy movement, sponsoring protests in Chicago, Toronto, London, Tokyo, Madrid, Milan, and Stockholm. The now omnipresent Guy Fawkes mask popularized by members of Anonymous (who call themselves Anons) can be seen at these protests worn by protesters who seek to hide their identities the way that the original Anons did during Project Chanology. The reasoning behind their actions is varied. Some espouse and adhere to an "activist" belief system, justifying their actions in reprisal to acts of greed or oppression. Others adopt the mantle of revolutionary freedom fighter. They seek to rid the world of the evils that have plagued it since before recorded history. Still others are looking for a cause; perhaps this is the most dangerous of all, as their desire to be a part of something can have serious repercussions. And still others are in it all for the lulz.

PROPAGANDA AND LULZ

There is something terrifying about the power of anonymity. People who believe that they are truly anonymous, or truly incapable of being identified regardless of who is looking for them, can be scary. Their actions often become unpredictable or, worse, pathologically predictable, taking full advantage of the perceived freedom that anonymity affords them. A great deal of research has been conducted in respect to this in behavioral psychological circles. The Proteus effect,[52] for example, suggests that it is possible (and likely) for an individual's behavior to conform to his or her digital self-representation. Put another way, the Proteus effect states that people's identity, how they carry themselves, their speech and interactions, can be influenced and directed by their online interactions and involvement in avatar- or handle-driven environments (collaborative virtual environments, multiuser domains, role playing environments, and so on). In some cases, the combination of the Proteus effect and perceived anonymity is intoxicating and proves irresistible to individuals who may otherwise be lacking or missing something from their lives. This, however, does not free a person from his or her obligations and responsibilities. People cannot simply do what they want just because they believe they are anonymous and beyond reproach. Such beliefs suggest a predisposition toward antisocial behavior and a genuine disregard for society.

Throughout this chapter, we have discussed several examples of individuals and groups who utilized anonymity as a means of protection and secrecy. Some of these groups espouse grandiose dogma and propaganda-driven rhetoric. Others are simply driven by seeing what they can get away with while hiding behind this thin veil of anonymity, not caring who they harm or what acts of unlawfulness they are accountable for in the process. They are simply in it for the lulz and that is enough for them, or so they say…

An offshoot of Anonymous, Lulz Security or LulzSec, came into being in 2011. The organization claims to be in it for the edification and improvement of security in its targets and has been highly critical of "whitehat" hackers who it claims have been corrupted by their employers.[53] LulzSec began its illegal activity in May 2011 when it targeted and breached the Fox.com Web site. Through this compromise, the LulzSec members, who stated they launched the attack due to the rapper known as Common being called vile on the air, gained unauthorized access to employee user IDs and passwords, LinkedIn accounts belonging to Fox.com employees (of which many were altered), and the personal information of more than 73,000 X-Factor contestants[54] which they leaked onto the Internet. Later escapades of the LulzSec hacker collective (whose greatest scores came through the exploit of SQL injection vulnerabilities in weakly secured Web infrastructures) included attacks against British banking ATM networks, and involvement in the Sony Japan hack where they claimed responsibility for leaking and publishing data mined from Sony databases.[5]

This would not be the last time that LulzSec exacted its brand of frontier justice against Sony. The group would go on to steal codes for music, coupons, and customer information. In an online rant, the group tweeted, "Everything we have will be published in multiple ways to ensure maximum embarrassment and exposure for Sony and their security flaws." The group would continue its online marauding through summer 2011, launching DDoS attacks against the CIA's publicly facing Web site. In June 2011, the group released a statement globally announcing its intent to partner with the Anonymous hacker collective in encouraging supporters to hack into, steal, and publish classified government information. This was an antagonistic move on the part of the LulzSec group and was deemed by many as a declaration of cyber guerilla warfare against governments and corporations which the group or its consorts held something against. It was a marked departure from the group's earlier activities, seeing them move more so into the realm of the "blackhat" than ever before.

Eventually, in late 2011, LulzSec departed as quickly as it had formed. The group released a public statement confirming its numbers and stating that the media had grown tired of the group and the group had grown tired of itself. But this wasn't entirely the end of LulzSec and its membership. Several members

---

[5]Ibid.

were thought to matriculate back into the larger, more nebulous, hacker collective of Anonymous. Law enforcement the world over also began to pursue and exercise arrest warrants against LulzSec members. Over time, Topiary[55] and Ryan McNeary[56] were arrested and taken into custody in Great Britain. Investigations into the remaining members of LulzSec continue at the time of this writing.

SUMMARY

Throughout this chapter, we examined a broad swath of phreaking and hacking history. We explored the birth of organized, underground hacking collectives and highlighted several though certainly not all that have or do exist. We explored the organizational structures, research projects, tools, and utilities that many of these organizations and their members have developed, some for the greater good and others for less honorable means. We also began the long journey of exploring the economic realities presented by these groups and groups like them around the world, of varying sophistication and skill. Understanding the mindset of individuals engaged in activity considered unlawful or illegal requires many skills, some learned, some possessed from birth. Understanding their motives requires patience, observational skills, access to data for the purpose of assembling a composite profile, and intelligence germane to the marketability of the skills, tools, and goods these organizations (and other, more sophisticated, professional criminal organizations) may possess. As you will see throughout the rest of this book, the potential for economic gain ranks quite highly among reasons to endeavor into this realm.

References

[1] Steinmeyer J. Hiding the Elephant. New York: Carroll & Graf Publishers; 2003.

[2] www.bibliotecapleyades.net/tesla/esp_tesla_18.htm.

[3] http://voyagerblog.com.au/2012/01/18/behind-the-maskelyne/.

[4] www.urbandictionary.com/define.php?term=lulz.

[5] http://science.hq.nasa.gov/kids/imagers/ems/consider.html.

[6] Hong S. Wireless: From Marconi's Black-Box to the Audion. Cambridge, MA: The MIT Press; 2010.

[7] http://ieee.ca/millennium/radio/radio_differences.html.

[8] www.history.com/this-day-in-history/marconi-sends-first-atlantic-wireless-transmission.

[9] www.hamradio.piatt.com/poldhu.htm.

[10] www.bbc.co.uk/history/worldwars/wwone/summary_01.shtml.

[11] http://necrometrics.com/20c5m.htm.

[12] www.u-s-history.com/pages/h1108.html.

[13] www.machinae.com/crypto/timelinepww2.html.

[14] www.ma.hw.ac.uk/~foss/valentin/Polish_breackdown.html.

[15] www.historyofinformation.com/index.php?id=1991.

[16] www.hums.canterbury.ac.nz/phil/people/personal_pages/jack_copeland/pub/etsamp.pdf.

[17] www.historyofphonephreaking.com/docs/weaver1954.pdf.

[18] www.historyofphonephreaking.org/docs/breen1960.pdf.

[19] http://blog.historyofphonephreaking.org/2008/08/wikipedia-an-information-ouroboros.html.

[20] www.historyofphonephreaking.org/docs/rosenbaum1971.pdf.

[21] http://bak.spc.org/dms/archive/britphrk.txt.

[22] www.textfiles.com/magazines/LOD/.

[23] http://www.google.com/url?sa=t&rct=j&q=leonard%20rose%20legion%20of%20doom%20indictment&source=web&cd=6&ved=0CDoQFjAF&url=http%3A%2F%2Fwww.docstoc.com%2Fdocs%2F87678943%2FThe-Secret-Service_-UUCP_and-The-Legion-of-Doom&ei=L-pGT8bHK5L3ggfOqYzrDQ&usg=AFQjCNH1z_e6U2NWql-agOLdOdJsW3uopQ.

[24] http://massis.lcs.mit.edu/archives/security-fraud/len.rose-legion.of.doom.

[25] www.cultdeadcow.com/tools/.

[26] http://w3.cultdeadcow.com/cms/apps.html#list.

[27] http://w3.cultdeadcow.com/cms/nsf.html.

[28] www.cbsnews.com/8301-201_162 57383882/bradley-manning-arraigned-but-enters-no-plea/.

[29] www.elsevierdirect.com/ISBN/9781597496131/Cybercrime-and-Espionage.

[30] www.4chan.org/.

[31] www.wired.com/techbiz/it/news/2004/07/64193.

[32] www.dmagazine.com/Home/D_Magazine/2011/April/How_Barrett_Brown_Helped_Overthrow_the_Government_of_Tunisia.aspx?p=1.

[33] www2.citypaper.com/columns/story.asp?id=15543.

[34] http://statenews.com/index.php/section/featuresindex.php/blog/entertainment/2008/02/internet_group__.

[35] www.computerworld.com.au/article/206359/anonymous_group_declares_online_war_scientology.

[36] www.telegraph.co.uk/technology/internet/8013548/Music-and-film-industry-websites-targeted-in-cyber-attacks.html.

[37] www.guardian.co.uk/world/2010/dec/31/anonymous-hackers-zimbabwe-wikileaks.

[38] www.britannica.com/EBchecked/topic/1753072/Jasmine-Revolution/299733/Time-line-Jasmine-Revolution.

[39] www.aljazeera.com/indepth/opinion/2011/02/201121321487750509.html.

[40] http://geohotgotsued.blogspot.com/.

[41] http://bcts.bergen.org/.

[42] http://cty.jhu.edu/.

[43] www.rit.edu/.

[44] www.purplera1n.com/.

[45] http://cydia.saurik.com/.

[46] www.limera1n.com/.

[47] http://news.bbc.co.uk/2/hi/technology/8478764.stm.

[48] http://psx-scene.com/forums/f6/geohot-here-your-ps3-root-key-now-hello-world-proof-74255/.

[49] www.washingtonpost.com/blogs/faster-forward/post/report-facebook-hires-playstation-hacker-george-hotz/2011/06/27/AGt0o1nH_blog.html.

[50] http://technologizer.com/2011/04/11/sony-george-hotz-settle-ps3-hacking-lawsuit/.

[51] www.adbusters.org/campaigns/occupywallstreet.

[52] http://vhil.stanford.edu/pubs/2009/yee-proteus-implications.pdf.

[53] http://online.wsj.com/article/BT-CO-20110605-702801.html.

[54] www.foxnews.com/scitech/2011/06/21/brief-history-lulzsec-hackers/.

[55] www.pcworld.com/businesscenter/article/236675/uk_police_arrest_anonymous_spokes-man_topiary.html.

[56] www.webcitation.org/5zdVEI7rE.

Drivers and Motives

INFORMATION IN THIS CHAPTER:

- Technology Advancements and Their Effect on Crime
- Motives for Committing Cybercrime
- Opportunistic Cybercrime Cost Model

INTRODUCTION

This chapter dives into the various dynamics that drive and motivate individuals to lead a life of cybercrime, along with the cost to organizations defending and protecting critical information. The primary objective of cybercriminals today is to obtain control, power, and wealth. Achieving this goal does not require geographic proximity to the targets these criminals want to control and exploit. Nor does it require that the cybercriminal be physically present to conduct the laundry list of activities associated with cybercrime today, including transportation from source countries of illegal narcotics, firearms, and durable goods, as well as the hacking of online payment systems, to name a few.

If you look back several decades, criminal organizations didn't have cell/satellite phones or the Internet to help them commit their crimes; they relied on hand-delivered mail, the telegraph, and eventually perhaps a fixed landline telephone. In addition, they were restricted to a geographic location, and almost every criminal exchange had to touch human hands, whether those hands belonged to a pawn, a mule, or the group's actual ringleader. Because of this, the risk of getting caught was much higher, and the evidence was not difficult to argue in a court of law, especially if the criminal was caught red-handed.

This is important to understand, as there have been some major shifts in how criminals are using advances in technology to minimize their risk of getting caught. In fact, advances in technology that connect us to work, family, and entertainment have created nefarious and dangerous underlying capabilities that are concealable and repeatable from almost anywhere on Earth. In fact, the chances of being caught commiting a crime on the backbone of the Internet are extremely low nowadays. This apparent global immunity introduces the increased propensity for any criminal organization (or person, for that matter) to exploit advances in technology, the Internet, and the people using it. Moreover, we now see that many traditional criminal organizations are venturing into cybercrime as a logical progression of their enterprising ways.

TECHNOLOGY ADVANCEMENTS AND THEIR EFFECT ON CRIME

The historical trends regarding advances in technology and their effect on crime are a key driver in the category of cybercrime. Thanks to higher connection speeds, cybercriminals have been able to expand their reach exponentially, and globally. No longer must we wait for a courier to deliver messages from afar. Today, those messages reach us instantly.

The reality of cybercrime is that it allows the perpetrator to break the fourth and fifth dimensions. Today's cybercriminals can cut across time and proximity, with blinding efficiency.

From Letters to Telegraph to Landline to Pager to Smartphone to SmartHome

The ever-improving communications backbone has provided great benefits to humankind, but it has also provided faster speed to criminals. Just like doctors can share patient x-rays as large files across hospitals, and even send them to other countries to be diagnosed overnight, so can criminals leverage the cyber backbone for their own advances. The unfortunate reality is that rarely do inventors consider the potential criminal intent that could be applied to their inventions, and therefore, rarely do they include security in their original designs. Security, when applied as an afterthought instead of being baked into the original recipe, will never be complete, and instead will be nothing more than icing on the cake.

Social Media and Location-Based Services

Social media and location-based services are extremely popular and almost as important as having an e-mail address or presence today. Facebook, Twitter, Foursquare, and Google Maps all have their purpose in connecting people at work, home, and play. However, Foursquare and other location-based services are like a Times Square-sized billboard that says, "I'm not home, please come and rob me blind!"

Indeed, traditional criminals are starting to use Twitter, Foursquare, Facebook, and Google Maps as tools to passively gather vital intelligence about their targets, without the targets even realizing they and their property are going to be victims in an upcoming crime. As a case in point, a recent SocialTimes article on a survey of burglars conducted in the United Kingdom stated that 78 percent of the 50 burglars surveyed used social media to plan their crimes.[1] Conversely, law enforcement officials have been very successful in using social media to track criminals that are on the run from the law. For instance, we recently interviewed a bounty hunter who explained that he will social-engineer his way into a subject's Friend network on Facebook, and based on photos the person posted containing house addresses and car license plates, among other things, he can provide the case investigator with enough information to run a trace to accurately locate the subject. Additionally, with the geo-tagging data that accompanies most photos people take nowadays, location-based information is also attached to these photos as metadata that can be used to geo-locate the subject to within a few feet of his or her actual location. Of course, this was not how Facebook was intended to be used. But just as cybercriminals exploit new technologies for bad, so can law enforcement and world governments use the same technologies to capture and/or monitor wanted individuals.

Datacenter to Desktop to Mobile Computing

As the volume and importance of data has migrated farther from the datacenter, that data has become increasingly vulnerable and harder to protect. During the era of mainframes, data administrators could protect a business's data with role-based access profiles and essentially hold company information "hostage" in a datacenter. Because businesses demanded data flexibility and mobility, security was rarely included in the original requirements of these datacenters, and businesses rarely considered the potential risk of losing their data.

Additionally, the popularity of bring your own device (BYOD) and the lack of mobile data management (MDM) within most enterprises further complicated the ability of organizations to control and protect their data; instead of data being centrally located, it was stored on a multitude of devices located both inside and outside the corporate infrastructure. Furthermore, the recent discovery of the malware Skywiper (a.k.a. Flame) has demonstrated that Bluetooth-enabled devices can be used as vectors to infect and transmit data. If this doesn't concern you when it comes to BYOD in your enterprise, it should at least make you take inventory of the "unknown" devices that are connected to your network. With that being said, traditionally the amount of money that is allocated toward security in most infrastructures is focused on corporate-owned assets and not BYOD, and therefore, any additional capital and operational expenses for securing BYOD are unlikely in most organizations' fiscal plans, at least for the next few years. As a result, cybercriminals will go after the lowest hanging fruit, and that shift will be seen in mobile devices.

eBay, Amazon, Craigslist, PayPal, and Online Offshore Financial Institutions

The online economy really started to take off in the mid-1990s, when many organizations began developing their online presence. According to the Digital Research Initiative, the birth of e-commerce took place August 11, 1994, with the first online transaction recorded by NetMarket.[2] The first item purchased via a Web site protected by commercially available data encryption technology was the CD "Ten Summoner's Tales" by Sting, according to NetMarket founder Daniel Kohn. One of Kohn's Swarthmore College classmates purchased the CD with his credit card for $12.48, plus shipping costs.

This set in motion the ability for any company to conduct business in a global economy 24 hours a day, seven days a week. Geographic boundaries did not matter; as long as you had a decent Internet connection, introduction to the marketplace was low-cost. Unfortunately, these very advances in technology introduced multiple opportunities for cybercriminals to distance themselves from their victims, and the virtual scene of the crime became 1s and 0s in the ether, causing great difficulty for the law enforcement community. And today, despite continuing advances in online tools and technologies, the digital marketplace remains a stomping ground for cybercriminals the world over.

MOTIVES FOR COMMITTING CYBERCRIME

Motives for committing organized cybercrime and state-sponsored cybercrime are similar in terms of the end goal. As we mentioned earlier in this chapter, perpetrators of organized crime are focused on control, power, and wealth. State-sponsored cybercrime is no different, as these criminals focus on control, power, and wealth at the national level instead of at a small group level. Wealth and control in the wrong hands ultimately leads to power. The right information in the wrong hands can result in anywhere from $10 to hundreds of millions of dollars. With this in mind, the following subsections discuss a few examples of the motives behind various groups of cybercriminals.

Organized Cybercrime

Many organized crime groups and crime families could be considered poster children for this relatively new era of cybercrime. The Russian Business Network (RBN) is just one example of a well-known organized crime syndicate that has leveraged the use of the Internet to conduct illegal operations. In his blog "Five Families of New York City," Dave Aitel stated the following: "If you think about the money that organized crime has, if they throw out $100,000 to attack you, it's hard for a corporation to fight against that."[3] The economics of fighting cybercrime, especially against a well-funded organization, might seem like a lost cause but with the right strategy and protections technologies in place you can significantly reduce you risk.

Ground Zero: The Eastern Block

All roads seem to point to Eastern Europe as the epicenter for a lot of organized cybercriminal activity. This became apparent to us after we researched several high-profile cybercrimes. In a few of the examples we provide in Chapter 9, you will notice that the key actors were reported to have lived in Romania and Eastern Europe, and the majority of those cybercriminals are still at large at the time of this writing. Media reports of terrorist groups such as Al Qaeda note that these groups allegedly have training bases in Pakistan and Afganistan. Have you ever wondered why Al Qaeda has training bases in those countries? The answer is, quite simply, that they are Third World countries that don't have a lot of control over their citizens, and that these citizens, regardless of their age, are looking for a common cause that will bring them together and make them feel important and socially connected within their religion. Furthermore, Eastern European countries have, for the most part, very weak cyber laws and security, unless they relate to the areas of banking or government. In fact, it has been reported that some Eastern European companies are only just installing firewalls and intrusion detection/prevention systems. This idea of treating security as an afterthought is beginning to change, but for the time being, for criminals who want to fly under the radar of local and federal law enforcement, Eastern Europe is probably their best bet.

Drug Trafficking and Organized Cybercrime Statistics

Statistics regarding drug trafficking and organized cybercrime are very telling; at the time of this writing, the organized cybercrime trade was valued at several billion dollars, while the drug trafficking trade was valued at around $1 trillion. This has been the cause of heated debate within the security community, as some experts believe the organized cybercrime trade is more highly valued than the drug trafficking trade. Despite the debate, the one fact regarding both of these illicit businesses is that you are less likely to get caught committing an organized cybercrime than you are trafficking drugs. Drug trafficking requires that the criminal physically make a transfer or transaction that involves human interaction. Organized cybercrime does not require the criminal to be physically present at the scene of the crime, thus making it very difficult for law enforcement officials to capture him or her. However, if the cybercrime is committed against a well-known organization such as CNN, Sony, or Hewlett-Packard, for instance, the criminal's chances of being caught are substantially higher than if he or she were conducting a breach and stealing information from a small business.

State-Sponsored Cybercrime

The first discovered instance of a government institution being successfully targeted and exploited by foreign nationals was the breach of Milnet by German nationals in the 1980s. The exploit was discovered and pursued by Cliff Stoll, who at the time was working as systems manager for the computer at Lawrence Berkeley National Laboratory, in Berkeley, California; after discovering a minor accounting error, Stoll became suspicious and eventually discovered that a hacker was using the Berkeley computer to hack into U.S. research and military computer networks. Stoll documented the case in his book, *The Cuckoo's Egg: Tracking a Spy Through the Maze of Computer Espionage.*

More recently, in 2012 the United States officially filed its first cybercrime charges against the Chinese government. The typical MO of a state-sponsored organization at the very highest level is to conduct esponiage, steal technology, and steal secrets. All of these are examples of crimes, and if carried out in most countries they would be considered treasonous.

Terrorism and Crime

Another example of cybercrime concerns terrorism. In recent testimony to the Senate Select Committee on Intelligence, FBI Director Robert Mueller said that "threats from cyber espionage, computer crime, and attacks on critical infrastructure will surpass terrorism as the number one threat facing the United States."[4] This comment didn't grab a lot of media headlines in the United States at the time, but it emphasized the fact that regardless of the motives of one individual or group (state/nonstate-sponsored actors/terrorists), cybercriminals will carry out their acts of theft, espionage, and sabotage with the use of interconnected devices on the Internet.

OPPORTUNISTIC CYBERCRIME COST MODEL

The barrier to entry in the cybercrime realm is relatively low. It is so low, in fact, that common street gangs are taking advantage of the opportunity to grab a piece of the $388 billion cybercrime trade, according to a recent study. The return on investment (ROI) for a traditional gang, common street criminal, or crime family to move from trading drugs to becoming involved in cybercrime can be compared to trading a penny for a dollar. In some countries, such as Vietnam, cybercriminals without their own Internet connection line up 20-deep for a crack at using cyber cafe stations to search for vulnerabilities in code that they can sell. Mind you, finding these vulnerabilities may not be illegal in most jurisdictions. Merely selling these vulnerabilities is not illegal either. In fact, some companies misguidedly choose to pay bounties for discovered vulnerabilities.

In addition, the weaponizing of threats to exploit these vulnerabilities may not be illegal in many jurisdictions. For example, the earliest worm, the Morris worm, was created in (and escaped from) a laboratory environment as an academic proof of concept.

Furthermore, selling these exploits online may not be illegal in many jurisdictions. However, use of these exploits may carry some form of criminal punishment, should you be caught and prosecuted, and the probabilities of that are infinitesimal.

The value chain of malware that is developed and the risk of prosecution do not grow linearly in terms of cost. Here is an example of the tools in a typical cybercrime kit, along with their associated costs:

- Laptop: $199.99 from www.pcexchange.com
- Wireless connection: free by using www.wififreespot.com/tex.html
- ZeuS Builder, a crimeware tool for building and configuring a ZeuS bot: $7,000
- Anonymous proxy service: $102.96 from http://provpnaccounts.com/ Buy_VPN_Account-118-articles

Total cost: $7302.95

Putting together a cybercrime kit does not always have to involve purchasing hacking software such as ZeuS Builder. The nefarious cybercriminal can also download freeware and free do-it-yourself (DIY) kits. However, the ROI for an investment of $7,302.95 (which most likely comes from stolen money to begin with) could be ~$6 million, which is what a German gang using ZeuS Builder reportedly netted. Based on the information in Table 3.1, it's not surprising that criminal organizations will shift their efforts to cybercrime.

This is only an example for a criminal entity that is just starting out, but if the cybercriminals have been operating for some time, it's likely that the gross percentage would be from 100 percent to 300 percent, as they would be reallocating funds that were stolen in the first place.

Table 3.1 Cybercriminal Profit Model			
Startup Cost	**Profit**	**GM**	**GM%**
$7,302.96	$6,000,000.00	$5,992,697.04	99.88%

Organizational Security ROI

One of the biggest questions that security professionals should ask is: "Am I doing the right things in order to protect the data that I am chartered with protecting?" This is not an easy question to answer, as it requires a depth of introspection rarely found in tactical practicioners. Unfortunately, those higher in the strategic hierarchy are rarely looking at the simple questions either. As such, we end up with a reactive and reactionary approach to security, where each symptom begets a new tool or appliance, without looking to solve the root cause.

The owners of the data, those that create it and multiply it, have left the stewardship of the data to the custodians in the datacenter. They have effectively outsourced the responsibility to the data minions (if you are reading this, you are probably one of said minions), without giving them authority to make any decisions over the data.

Because of this disruptive lack of discipline in security, the measure of a security "return on investment" becomes a laughable concept to most business owners. They see the security department as a black hole that opens once a year to ask for inordinate amounts of money, and then disappears for the rest of the year. If nothing goes wrong, they are happy to take credit for the silence, but if something goes wrong, they literally add insult to injury by requesting additional money to make up for the fallacies in their previous approach, thus betting on the fact that all the additional tools will lower the probability of that particular failure happening again in the foreseeable future.

If an organization is willing to truly consider looking for a return on its investment in security, often the best approach is to measure the reduction in risk as a return on security investment.

SUMMARY

In this chapter, we discussed how advancements in hardware and software have been and will continue to be the primary driver/catalyst of cybercrime. The Internet is becoming faster, bigger, and more agile in an effort to meet our demands for working, socializing, living, and playing online. With bigger Internet pipes and palm-sized smart devices, cybercriminals can conduct their operations with the click of a button, thereby capturing terabytes of data. The key motive for these cybercriminals, regardless of the entity they support, is to steal data that has value. The value of this data can be measured on different levels based on what the cybercriminal plans to do with it. For example, organizations that are state-sponsored are often silent and persistent, and will go to great lengths to ensure that they have access to vital corporate and government information. The scary aspect of this real-life scenario is that before the compromised target of state-sponsored activity is made aware of the breach, the cybercriminals have already stolen terabytes of data. The data that has been captured can net the state sponsors advancements in technology, military, and intelligence planning activities that provide them with unfair competitive advantages. The monetary value of such information can be in the hundreds of millions of dollars.

The key point to take away from all this is that although advancements in technology will continue to drive the way we communicate on a daily basis, they will also provide nefarious cybercriminals with new methods for exploiting information.

References

[1] http://socialtimes.com/new-survery-burglars-use-social-media-to-plan-crimes_b79475.

[2] http://news.cnet.com/E-commerce-turns-10/2100-1023_3-5304683.html.

[3] www.fivefamiliesnyc.com/2011/07/russian-mob-runs-world-of-internet.html.

[4] http://abcnews.go.com/blogs/politics/2012/01/fbi-director-says-cyberthreat-will-surpass-threat-from-terrorists/.

CHAPTER 4

Signal-to-Noise Ratio

INFORMATION IN THIS CHAPTER:

- Cyber Attacks: The Early Years
- The Pendulum Swings Back: Hacktivism and DDoS

CONTENTS

INTRODUCTION

Signal-to-noise ratio (often abbreviated as SNR or S/N) is a measure used in science and engineering that compares the level of a desired signal to the level of background noise. The signal-to-noise ratio, the bandwidth, and the channel capacity of a communication channel are connected by the Shannon–Hartley theorem[1]. Signal-to-noise ratio is sometimes used informally to refer to the ratio of useful information to false or irrelevant data in a conversation or exchange. For example, in online discussion forums and other online communities, off-topic posts and spam are regarded as "noise" that interferes with the "signal" of appropriate discussion. When dealing with malware, the size and stealthiness of the malware is a key piece in its potential success. If malware can inject itself into a broader stream of communications, it becomes harder to detect. Likewise, if the outbound fruits of successful malware, be it passwords, information, or files, can be injected into outbound web surfing streams, the impact becomes harder to detect as well. In this chapter, we will discuss the development of malware, and how it has maintained a low signal-to-noise ratio to improve its efficiency.

---

[1] An explanation of the Shannon Hartley Theorem applied to data streams is included in a white paper published by MIT and IBM that can be found at http://www.almaden.ibm.com/cs/people/dpwoodru/pw12.pdf.

CYBER ATTACKS: THE EARLY YEARS

As we discussed earlier in this book, the concept of a virus originally was intended to be more of a claim to fame, an actual ode to the hacker in the truest sense of the word. In the first viruses, such as Creeper, which was released in the Advanced Research Projects Agency Network (ARPANET) in 1971, a message was displayed on the infected system. The message made it clear that there had been an infection, and often, who had perpetrated it.

Later, viruses looked to celebrate a specific event, to a certain extent. The malicious payload of the Michelangelo virus, for instance, was only triggered on the birthdate of the renaissance artist, remaining dormant in the system for weeks or months prior to its release. There was a certain artistic provenance regarding the thoughtfulness of this virus's developers. I'm not sure that it was a critical global issue that people did not realize Michelangelo was born on March 6, but the virus definitely called attention to that specific date in 1990 by celebrating the artist's birthday through mayhem.[2] The media gave attention to the issue because nothing like the Michelangelo virus had been seen before. It was, literally, newsworthy and intellectually motivated, and despite the potential damage it caused to the average computer user, all data on the drive was respectfully maintained. It only impacted the boot sector of the hard drive or floppy.

Perpetrators back in the 1980s were looking to highlight their skills and mastery over the computer operating system and applications. They could co-opt behavior from the system beyond the design specifications, and that made them proud of their accomplishments. Some of the more benignly intentioned hackers were looking to correct what they saw as flaws in design or function, and intended to pressure vendors into correcting these errors. They hunted for bugs and sought to improve the ecosytem, even if their ways to the means were questionable. Others were merely looking to leverage vulnerabilities in the system to draw attention to themselves.

In a way, when viruses were more interested in propagating merely to do harm to operating systems, they were more like Lindsay Lohan, acting out merely to be noticed, without causing real harm to others, except for the odd accusation of shoplifting. They came in with a big brass band, almost immediately notifying the infected target of their presence by flooding the target with information, changing boot sectors to prevent the system from returning after a reboot, or changing wallpapers and colors on the screen. They were nuisance viruses, but they were easily identified, as the main motivator of the malware

---

[2] A clear explanation by F-Prot of the Michelangelo virus can be found at http://www.f-prot.com/virusinfo/descriptions/michhelangelo.html

developer was to gain instant personal notoriety. In fact, many of these viruses, as well as Web page defacements, were literally signed by the perpetrators, in the same way that artists sign their masterpieces. As well, because of their simple intent, the correction was often just as simple: Rewrite the boot sector, as with the Michelangelo virus, and you regain access to your fully operational system. Merely copy back the file index.html, and your Web page is magically back. Copy win.ini back from a location where it was thoughtfully left behind by the hacker, and your system is automatically back in business.

As the main motivator has now shifted from fame to fortune, however, the focus has shifted to stealth, and to losing your hacking self within the system's inherent noise. No drums, brass bands, and cymbals are wanted now, because they only attract undue attention. Simply stated, if a cybercriminal can create an ATM-specific Trojan and stay in the system undetected for one day, he or she may be able to collect 50 credit card numbers and their respective PINs. If the cybercriminal can remain undetected in the ATM for a week, we are talking about 400 credit cards and PINs. After a full year, we are talking about a complete revenue stream that is easy to monetize and extremely difficult to prosecute. With the amount of data that is being transferred in each transaction, the purloined data becomes a very small amount of noise to detect within the system's signal.

This pattern becomes increasingly common as hackers look to monetize their crimes. The lower their criminal noise can be in relation to the system's valid signal traffic, the longer they are likely to perpetrate the crime undetected. Because of this motivation, they investigate, and often resort to purchasing, "zero-day" vulnerabilities whose potential heat signatures are not yet known to the anti-malware community. Intelligent attackers seek to hide their behavior by obfuscating and packing their messages, and they invest heavily in development cycles to make their state-of-the-art malware more difficult to detect and remove. We see million-dollar development efforts in some of the more sophisticated and dangerous malware examples. The bad guys are investing these additional resources only make it harder to defend against their malware, and to underline the fact that this is now a viable business for them. They are no longer looking to hit the corner liquor store for $50. They are looking to embed themselves much deeper into the ecosystem, often well in advance of when they will exploit the vulnerability.

A clear, insidious, and dangerous example of this can be found in the Stuxnet attacks, now known to have been perpetrated by unfriendly countries against Iran's nuclear infrastructure, where we see a new use for malware as a laser-targeted weapon of an undeclared and potentially illegal cyber war. The developers of the malware sought to insert themselves into a specific brand of programmable logic controllers (PLCs) that were likely to be used in a nuclear reactor's centrifuge. Interestingly enough, not all of the PLCs were susceptible to the attack. A vaccine was included in the malware for those PLCs in friendly geographic areas. The compromised PLCs were mostly found in countries that were like-minded to Iran, such as Venezuela and Ecuador, but not in Colombia, for example (see Figure 4.1).

These PLCs were reconfigured to report back normal settings, while they were accelerating the hardware to the point where they actually caused damage to the centrifuge infrastructure. The operators were dumbfounded because their dashboards all showed readings that were within the expected parameters, but the centrifuge was being damaged. This kind of malice aforethought goes well beyond the opportunistic hackers of yore and into a much deeper and more dangerous model of cyber warfare that requires investment levels far beyond those of typical script kiddies. In the case of Stuxnet and Duqu, we now have clear and concise evidence that these weapons were developed by the United States government, and that they were deployed deliberately, if irresponsibly, upon the global interconnected community. Moreover, the government's admitting to having released those cyber weapons has quietly unleashed a new era of global war, where escalation will yield mutually assured destruction, a concept not considered seriously since the 1980s at the height of the nuclear arms race. Pandora's Cyber Box is now open for business. Countries will now

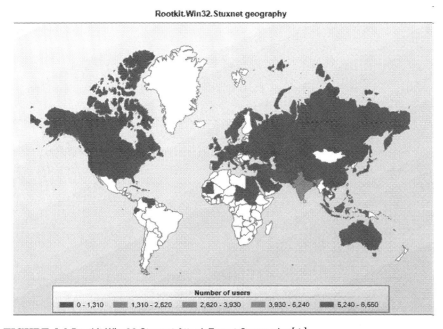

FIGURE 4.1 Rootkit.Win.32.Stuxnet Attack Target Geography [1]

be forced to develop offensive cyber war capabilities, in essence making them compete with cybercriminals for hacking resources.

Interestingly, we now see a common infrastructure, at an intimate level, among three of the most insidious known cyber weapons: Flame, Stuxnet, and Duqu. According to research performed by Kaspersky Lab, common code is included in all three of these weapons, which seems to imply, despite very different attack vectors and manifestations, that the developers of these weapons shared common tools and development methdologies. Flame is a behemoth of a weapon, coming in at a hefty 20 MB when one includes various plug-ins that are available, and having deployed quietly, despite its heft, for the past two years across the cyber world. It includes a packed module that was later found in Stuxnet, although the module is unpacked in Stuxnet. This could be based on the need for agility in Stuxnet's target systems, PLCs. The size of the delivery package allowed Flame to carry multiple spy weapons within it, including the capability to record with the PC's microphone, to capture screenshots, log keystrokes, and send this data to a series of hosted domains that were all created bogusly. The code could be updated remotely, and the list of command and control domains could be updated on the fly. It was not a particulary beautiful piece of code in terms of efficiency, as it is believed to have been developed by multiple teams modularly. Team A, in charge of deployment, had no knowledge of the payload. Team B, in charge of finding and exploiting zero-day vulnerabilities as entry vectors, had no understanding of the propagation methodologies. Team C, in charge of the actual spyware payload, had no knowledge of how or where it would be deployed. Team D, in charge of delivering the purloined information to a series of repositories, was not responsible for the registration of domains or the deployment of the actual data capture servers. Team E, in charge of stealth data collection and actual spying, had no idea how the data was collected or that malware had been involved.

Looking at it from a historical framework, Flame had to predate Stuxnet, since it acted as the reconnaissance team, going deep early and gathering intelligence for the future attacks. Flame is considered by many to be nuisance spyware, and this is, in many ways, a correct assessment of the weapon. However, when considered as part of a government-sponsored effort with a development budget in excess of $1.2 million, along with how targeted it was in terms of deployment and focus, the threat increases exponentially for the intended targets. The rest of the world was not threatened directly in the same ways that North Korean tanks pointed at South Korea are not a threat to China. However, if these weapons, be they Flame, Stuxnet, or a tank, are redeployed to point at anyone else, the imminent threat becomes real to the new target. These laser-focused attacks, with such deep development pockets, are bound to be successful and, like other weapons, are likely to be misused and obtained by entities that will use them in an irresponsible manner. The reality of asymmetric cyber warfare is now upon us. How we, as a world community, react to this new reality will dictate whether we develop further technologies, or sink back into a romantic era of candle-lit dinners due to the destruction of the electrical infrastructure by some SCADA-savvy cyber weapon gone awry. One well-placed cluster of grapeshot can definitely have extremely dangerous repercussions on many unintended targets. The unintended consequences of such an attack should not be minimized. We are now involved in a Cyber Cold War, and we need responsible thought leaders to protect us through intelligent use of weapons, global frameworks, and nonproliferation treaties. The concept of these new hyper viruses in the hands of unfriendly nongovernment organizations should create a visceral reaction on every potential target. The closeness of a potential "Digital Pearl Harbor" should be a concern on all of our minds.

The fact that command and control communications could be hidden within the normal traffic for those devices also speaks to the sophistication of their development. The malware was designed and developed with a keen focus on minimizing its impact on available resources. A PLC, for all intents and purposes, can be considered a dumb device by today's standards. A PLC, much like the tiny brain that allows the windshield wipers in a car to operate at multiple speeds and intermittently, is limited by its design parameters and is intended for a particular purpose, and as such, we would not expect that same PLC to easily go beyond its defined parameters. In this particular case, a foreign entity was able to remotely reconfigure these PLCs, and still manage to remain undetected for months, which speaks to extremely advanced programming and quality assurance, factors that are rarely found in "normal viruses." In other words, the signal ratio of today's advanced malware ecosystem is much lower, and easily hidden within normal traffic noise.

The net result of Stuxnet is that 20 percent of the Iranian centrifuges were physically damaged under the very watchful eye of the facility's monitors. Their only grievous sin was to trust their monitoring equipment, and not realize that their infrastructure had been severely infiltrated and compromised. Like in every *Mission Impossible* film, an effective false facade was deployed to fool the sentries, while foreign entities, unfriendly to their efforts, had managed to deliver an incredibly efficient, highly stealthy, and incredibly effective weapon. The element of surprise could not have been greater. The plot was not new, but the theater of operation was.

The latest cyber weapon to be discovered that has similarities to Stuxnet, Duqu, and Flame has been dubbed Gauss by researchers at Kaspersky Lab. The scary aspect of this malware, first deployed in September 2011 and discovered in June 2012, is that it may actually be the first documented use of a government-grade cyber weapon, repurposed for cybercriminal deeds as a Banking Trojan.[2] This apparent code-cousin of Stuxnet and Flame is aimed at stealing personal

> The reason I say this is because Kaspersky Labs found this code and once the code is brought out of the wild, it can be deconstructed and sent back into the wild targeting the sender. It's like capturing a live Tomahawk missile and reprogramming it to return home and explode.
>
> While nobody can actually capture a live flying Tomahawk missile and do that, it's not impossible with computer code. This is more like capturing a Tomahawk, making 10,000 copies, and reprogramming them all to return home and explode. The United States will end up becoming the target of the attacks thanks to its own code.
>
> John C. Dvorak

information, specifically banking information, but leverages some of the same geographic controls of previous versions, including targeting machines in specific time zones. While Flame attacked mostly Iraqi address space, Gauss seems to be more focused on Lebanon.

Different modules of Gauss serve the purpose of collecting information from end users' Internet browsers, including the history of visited Web sites and passwords. Additionally, data on infected machines is sent to the attackers, including specifics regarding network interfaces, the computer's drives, and BIOS information. Lastly, the Gauss module is also capable of stealing data from the clients of several Lebanese banks, including the Bank of Beirut, EBLF, BlomBank, ByblosBank, FransaBank, and Credit Libanais, as well as specifically targeting users of Citibank and PayPal.

According to Kaspersky Lab, there are "strong resemblances and correlations between Flame and Gauss".[3] This type of activity is more aligned to cyber-crime than it is to cyber espionage or cyber terrorism. As such, it would appear that this type of cyber threat has leveraged the infrastructure of a government-sponsored super-malware, and has reverse-engineered the code so that it can be aimed at normal users. Unlike other weapons of war, when code is used to create a cyber weapon of these proportions, the code is sent over open channels. As such, mere mortals can intercept the code. When the code can be intercepted, even when it is encrypted and packed, it can easily be reverse-engineered with tools available online by any hacker with time and initiative.

At the risk of sounding like an alarmist, the possibility of this type of cyber weapon having its payload altered by a relatively skilled hacker presents a nearly incredible hazard to society at large. A powerful cyber weapon such as Stuxnet attacking Internet banking transactions is a definite risk. However, leveraging it to attack critical infrastructure through SCADA systems creates a much greater danger. With the knowledge that Stuxnet already had to attack PLCs, the risk to all critical infrastructures by subtle reprogramming of the payload is potentially catastrophic. The potential commercialization of such weapons, as intimated by John Dvorak, will create a rather disturbing problem for those that originally created the cyber weapons, and couldn't manage to harness their power once it was unleashed. "At the end of the day, there will be a government hearing and questions will be asked as to why this code was released in the first place. There will be no good answers."[2]

Using Stealth As a Weapon

Encryption started out as a weapon of war. The need to send information unknown to the enemy was critical in the development of the first ciphers. As far back as the 7th century BC, messengers delivered transport-encrypted messages to generals that were wrapped around a rod of wood of a very particular diameter. The receiver of the message, by using an identical diameter dowel, could easily read the message. This method of encryption, called the Scytale cipher, was first utilized by the Spartans and the ancient Greeks to transport information during battles.

Encryption, to this date, remains classified as a dual-use technology, and certain controls are still in place for the export of those technologies.'[4]

Polymorphism, Packing, and Encryption

As malware became more widely known, and anti-virus programs became more capable of detecting malware through patterns, the criminal element found a need to make these programs harder to identify as they attempted to enter target systems.

They started using multiple forms of hiding, in order to make it more difficult for the anti-malware programs to detect them. One of the first changes implemented to try to subvert anti-malware programs was polymorphism. In a polymorphic virus, each new iteration of the malware takes on a new characteristic, without impacting the main code. As such, it becomes harder to identify it with simple pattern matching.

By packing and encrypting the malware, cybercriminals escalated the arms race once again. With these techniques, they were often able to bypass base detection. These were the days of "Pray and Spray," when there was little targeting being done by attackers, and they mainly looked to reach the largest possible attack surface.

The packed and encrypted payloads forced the smart anti-malware providers to migrate to a heuristic engine so that the malware behavior could be detected, regardless of the path it took to reach the system.

The Need for Hierarchical Frameworks in Malware

The first botnet found in the wild was Bagle, discovered in 2004. Botnets differ from worms in their intent and use. While a worm looks to grow and expand through contact in a sort of "mine is bigger than yours" contest, a botnet actually installs command and control channels so that the bot herder can use the infected and compromised systems for a specific purpose, such as a denial of service or click attack intended to drive up their competition's cost of sales. Today, botnet machines are increasingly being used to lease attack surfaces against specific targets, instead of trying to pilfer money from innocent victims. In a very evolutionary fashion, cybercriminals have decided that there is less risk in leasing the tools for attacking than in trying to steal directly from their victims. Bagle actually had an integrated SMTP engine, which allowed it to convert any compromised machine into a mail server so that it could spam other users.

With more stable attack platforms, the command and control channel requires less redundancy and self-healing features to be included. As such, the amount of unecessary chatter between systems can be cut down substantially. When the attack platform was made up of slow and unstable dial-up connections, IP addresses changed constantly in the compromised systems, and a lot of CRC checks and redundancy were required to ensure the control of the attack platform was maintained. As such, the best way to manage a large botnet was through multiple bot herders, controlled by bot masters that leveraged multiple layers of hierarchy to grow the systems much like a military battalion has numerous divisions.

The Impact of Broadband

As broadband connectivity expands and the stability of attack platforms improves drastically, attackers quickly realize they can have the same impact with 20,000 compromised broadband users as they could previously with 400,000 unstable dial-up users. When network connectivity reaches universities, we see that now the same attack can be perpetrated with 2,000 strategically placed machines with ample bandwidth. As such, the logistical traffic required is substantially less, and the signal aspect of the attack versus the noise of management improves dramatically. In this case, the attackers essentially did the same thing as corporate America: They eliminated middle management posts. Now, with very few super nodes, strategically injected into compromised systems with ample bandwidth, the size of modern botnets can grow to hundreds of thousands of machines, creating an attack potential that can easily erode the best of connectivities, as exemplified by the Anonymous attacks discussed in Chapter 3. It is worthwhile to point out that many of the Anonymous attack systems were not technically compromised. Ideologically minded people volunteered their systems, as well as their botnets, to attack a common enemy.

As we move from cybercrime to cyber warfare, we see that the same tools the criminals used, when orchestrated and massified, can become a weaponized and powerful force for attacks between countries. In the same ways that the first knives yielded the swords of battle and the first guns begat rifles and tanks, we now arrive at a crucial crossroads in mankind's advancement. Will Stuxnet be the opening salvo in a cyber escalation that will only take us to mutually assured destruction? Or will it be, like the lessons of nuclear war, a weapon used once and then held back, due to the fears of unleashing it on humankind?

As we move from cyber war to cyber terrorism, and specifically, to state-sponsored cyber terrorism, the need for stealth deployment and configuration becomes critical to the success of these targeted attacks. In the same ways that missing weaponry from armies becomes the tool of the trade for cyber weapons dealers, they will now have super-cyber weapons in their arsenal to offer to the highest bidder, who without the proper moral compass can achieve great damage in assymetric battle. In the hands of a group that does not respect international treaties and conventions, cyber weapons such as Stuxnet can take us back to the Stone Age, destroying the very infrastructure that we rely on for water, electricity, and communications, in the blink of an eye.

THE PENDULUM SWINGS BACK: HACKTIVISM AND DDOS

When attackers break the surface nowadays, it is strictly to send a message, usually political. They no longer seek "geek notoriety"; rather, they have elevated their calling into a "cause notoriety". They look to make their way from the technology section on page 26, to the front page, and hopefully, above the fold with their attacks. The attacks that we hear about have impact well beyond the system that was compromised. We rarely hear about the individual machines that are compromised. We hear of the impact when these machines, acting as a collective, are put into action. The notoriety of the modern day hacktivist with a distributed denial of service (DDoS) attack that is called, much in the way Babe Ruth called where his home run would go, has the subtlety of a caveman. By harnessing the computing power of a botnet into a brute force bandwidth attrition attack of a server-saturation denial of service, cybercriminals use a force that could be used for extreme good, to draw attention to their cause. In this case, the message on the server is nonexistent, since the entire purpose of their attack is to take a politically motivated message off the radar. By taking a governmental or religious entity's server offline, they consider the message delivered. And the worst part of it is that they consistently succeed. Because of the asymmetric nature of the attack, and the vagueness of the targets, they often appear to be even more successful, much like a carnival fortune teller. Because the press and the public are looking to find success in their hacktivism, they often find it even when it is only circumstantial. The hacktivists often will present multiple potential targets, and if one is unassailable, they merely ignore the fact that they were unsuccessful and move on to a different target.

As hacktivism has increased, we again see a return to the target botnets. Because the notoriety of the previous attacks begat a technology response to the symptom in the form of anti-DDoS devices and services, larger and larger botnets were required to achieve the same results. The largest botnet in modern day hacktivism was allegedly OpMegaupload, which was controlled by the amorphous group Anonymous. On January 2012, the group used its Low Orbit Ion Canon (LOIC),[5] a tool that could be downloaded by those in agreement with them politically that allowed their computing resources to be used to target specific websites with Distributed Denial-of-Service attacks. The LOIC was used to bring down multiple U.S. government Web sites, including those belonging to the Department of Justice and the FBI, as well as Universal Music, the Recording Industry Association of America (RIAA), the Motion Picture Association of America (MPAA), and the U.S. Copyright Office.[6]

The last page in the history of this alteration, oddly enough, comes with the creation of separate private networks, segmented from the backbone of the Internet, for key ministries of certain countries such as Iran, "to shield them behind a secure computer wall from disruptive cyber attacks like the Stuxnet and Flame viruses."[7]

SUMMARY

Higher minded hackers, and lately, higher minded governments have learned that they need to substantially lower their signal to noise ratio to have successful cyber attacks. By hiding attacks within in the multiple data streams, they can lower their signal-to-noise ratios, and run their malware undetected for years.

References

[1] Courtesy of Symantec, presented in a lecture by Vitaly Shmatikov at University of Texas, Austin, http://www.cs.utexas.edu/~shmat/courses/cs378/stuxnet.ppt.

[2] Dvorak, John. Found at http://www.pcmag.com/article2/0,2817,2408307,00.asp.

[3] http://usa.kaspersky.com/about-us/press-center/press-releases/kaspersky-lab-discovers-%E2%80%98gauss%E2%80%99-%E2%80%93-new-complex-cyber-threat-desi.

[4] http://www.wassenaar.org/publicdocuments/index_CL.html.

[5] Johnson, Joel, Gizmodo – Found at http://gizmodo.com/5709630/what-is-loic.

[6] Peckham, Matt. TIME Techland. http://techland.time.com/2012/01/20/10-sites-skewered-by-anonymous-including-fbi-doj-u-s-copyright-office/.

[7] Tait, Robert. The Telegraph. http://www.telegraph.co.uk/news/worldnews/middleeast/iran/9453905/Iranian-state-goes-offline-to-dodge-cyber-attacks.html.

Execution

INFORMATION IN THIS CHAPTER:

- How Cybercriminals Execute Their Schemes Using Malicious Code and Content for Profit
- Identifying the Market
- Identifying the Target Audience

INTRODUCTION

In the first four chapters of this book, we investigated concepts dealing with psychology, history, motive, and patterns of behavior among cybercriminals. In this chapter, we will explore how cybercriminals execute their attacks. What is required for the successful execution of a cybercriminal operation? Is there any difference in planning and execution between this type of criminal operation and what you might see in non-cyber-related criminal initiatives? Are there differences between amateur and professional operations, and if so, what are they? By analyzing tactics, techniques, and procedures, can we establish profiles of these threat actors, criminal or otherwise, and dossiers that provide a clear and reasonably irrefutable assessment of the identities of those who are architecting and masterminding these operations?

In this chapter, we will examine models employed by cybercriminals around the world. We will consider their maturity while also studying their ability to operate skillfully so as to avoid drawing unwanted attention from their victims while in many cases demonstrating their ability to market their skills and wares professionally to broad underground audiences. This chapter will also provide insight into what could be considered the "go to market" strategy of cybercriminals; you will see, through examples, that this strategy is similar from those you may encounter when reading about product and service line management in modern IT infrastructure books or periodicals. Our goal is to provide a fresh insight to readers who have not previously been exposed to this aspect of the cybercriminal underground while at the same time providing detailed examples of how critical solid execution is to the success or failure of these enterprises. Our primary concern is that this chapter not be confused with a case-by-case analysis of examples seen in the underground, and instead, that it is viewed as a synopsis of the models, maturity, thoughtfulness (or lack thereof), marketing ability, and attention to detail employed by those who execute these strategies.

HOW CYBERCRIMINALS EXECUTE THEIR SCHEMES USING MALICIOUS CODE AND CONTENT FOR PROFIT

Members of the cybercriminal underground (proprietors) obviously conduct a large portion of their business—the selling of goods and services to interested third parties—on the Internet. However, this is just one aspect of the underground, and it does not represent all the actions that take place every day around the world. You will find that cybercriminals, like any other businesspeople, adhere to rules, customs, traditions, and indeed, the culture of global business.

All businesspeople understand the principle of supply and demand, whether their goal is to penetrate the online gaming market with a next-generation game system that offers advanced graphics and tactile game play, develop and market a new drug designed to treat erectile dysfunction, or promote the acquisition of their competitors' intellectual property by creating targeted malware delivered via a polymorphic botnet. This principle is neither new nor, as we've just seen, unique to legitimate business theorists. Understanding what is in demand and being able to supply and satiate that need is crucial to all successful businesses. As we will discover, being aware of the market conditions that fuel demand is just as important for the criminal as it is for the legitimate businessperson.

It is important to recognize that professional cybercriminals run their businesses as profit-seeking entities. They struggle with timelines, research and development, quality control, and distribution issues. They seek to maximize margins and minimize expense and risk. They gather business intelligence and engage in marketing and competition. Above all, they seek to make money. They do this through processes instantly recognizable to anyone in the world of software manufacturing.

From a technical standpoint, the economics are clear: Cybercriminals produce and then sell or use a software product that brings in more money than it cost to develop. For organized gangs engaging in the cybercrime equivalent of trading on their own account, this means, at the very least:

- Efficiently identifying a market (that is, enumerating targets that have money or salable goods in digital form and a vulnerability that can be exploited).
- Efficiently creating a tool to harvest the targeted data while preserving an acceptable level of anonymity.
- Ensuring that the workflow and processes enabled by the tool create or support the criminals' ability to commingle, launder, and access the ill-gotten gains.
- Continuing to develop and maintain the tool so that it meets all the requirements of the user at a profitable level.

Effective cybercriminals will take their time enumerating environments looking for vulnerable (susceptible) hosts that they can exploit in order to comprise in pursuit of a greater ends. That end will vary among cybercriminals based on their personal and collective goals, but the dynamics of this research remain constant. Other constants concern market intelligence, competitive intelligence, and counter surveillance of law enforcement activity in areas of direct concern—in other words, risk analysis.

It is also important to realize that these cybercriminals are running a business in all respects, including culturally. We have met people who happily state that they're malware engineers. This is not a hive of people furtively going about something; many professional cybercrime operations are run as, well, professional operations. An analysis of worker entry and exit from an operation like this will look identical to any other business: Workers arrive in the morning, take smoking breaks, take meal breaks, and finish their day in a predictable, business-like manner.

We mention this because understanding the realities of a cybercriminal operation requires understanding the totality of its conditions. From these conditions we can infer motive, efficiency, and efficacy, and we can trace observed shifts in focus to shifts in targets, activity, tactics, or procedures.

From a technical standpoint, there are countless ways to exploit computer, application, or other vulnerabilities for money—so many ways, in fact, that we won't attempt to list them all here. The economics of this, however, are relatively static, so in this chapter we discuss methods that we feel exemplify these economic principles. As we tell enterprise customers, the only adversaries you need to worry about are those who have figured out that, by spending $5 million, they can steal information worth $1 billion—in other words, the risk to organizations is about the economics, not any particular threat. There is, of course, a highly interesting line of discussion that leads from there into the best ways to defend against the tools that are the manifestation of these risks, but that is outside the scope of this book.

As a reasonably universal example of these manifestations, let's use commercial Trojans as the example of a typical cybercriminal tool used to extract data that can be converted to money. A Trojan is a software program that claims to do one thing, such as examine your computer for malware, or be a document or Web page you're interested in, sent apparently by a colleague, friend, or family member, but in fact does another. The name obviously derives from the Trojan Horse which, according to the *Aeneid of Virgil*, was used by the Greeks to invade the city of Troy during the Trojan War.

The basic workflow of this tool is as follows:

- Get the victim to execute the software (usually a link in an e-mail or on a Web site).
- Use the initial software run to identify the computer on which the Trojan is running.
- Communicate with a command and control server to seek further instructions and supplementary software.
- Run supplementary software and capture desired data (for example, passwords/usernames, account numbers, documents, and so on).
  - This step may also include additional steps to ensure that the infected computer is running well by cleaning up other malware, and other administrative tasks.
- Transmit (often after encryption) harvested data to a command and control server on a periodic basis.

As you can see, this is a simple model that has been repeated for the past several years. At each stage there is room for innovation or incremental improvement, however. For example, the method of communicating with the command and control server can be improved through new ways of randomly generating the list of servers with which the Trojan seeks to communicate, as well as more efficient methods of encrypted communication. How the malware attempts to thwart intrusion detection and prevention systems is another key area of innovation and improvement, as is how the Trojan behaves on the infected host to minimize interaction with host-based anti-malware and user complaints of poor performance.

When analyzing the profitability of any given campaign, then, you must take into account the costs of these innovations and incremental improvements, version upgrades and distribution, research and development of new exploits, and hosting and development of encryption, stealth, and similar activities, as well as the costs associated with launching the campaign, including victim research for targeted attacks (what would make a user click on a given document, and so on).

IDENTIFYING THE MARKET

In the preceding section, we outlined the economic activities of cybercriminal software developers who are basically trading on their own account. Like razor sellers who sell the handle cheaply and the blades expensively, and like those who sold picks and shovels to gold prospectors, cybercriminals find that the market for selling tools to those who would actually use them to criminally exploit computers is easier, less risky, and often more profitable than conducting the crimes themselves. For the cybercriminal developer, then, the task of identifying demand and being able to broadly cater to that demand is an attractive proposition.

There are ample customers to market to through a variety of channels, all of which are willing to pay a competitive market price for the good or service to be delivered by these crafty cybercriminals. Nevertheless, it would be foolish for an informed cybercriminal entity or organization to assume that *they* were the purveyors of fine criminal goods and services tapping into the global marketplace and that others (equally if not more enterprising than themselves) were not willing to go to extremes to secure the business of those driving the demand.

IDENTIFYING THE TARGET AUDIENCE

Assuming that the cybercriminal has identified the market and produced, procured, and packaged his or her goods and/or services, the cybercriminal then takes steps to identify the target audience. You may be wondering how this occurs within underground, black market economies. In many respects, target audience identification and acquisition among cybercriminals occurs in ways very similar to those in legitimate business. Amateurs and professionals alike rely on online advertising, but in well-controlled, vetted forums and underground criminal ecosystems. Figure 5.1 is an example of a simple yet effective form of underground forum-driven advertisement. You'll note that this cybercriminal offers three options. The first option offers prospective clients the following features for $1,000:

■ ZeuS 2.0.8.9 binary[1]
■ VNC (virtual network connection) remote connectivity tool.

---

[1] Originally supposed to have been sold to Harderman/Gribodemon, which we know occurred, but was the end of ZeuS outside and beyond the realm of SpyEye.

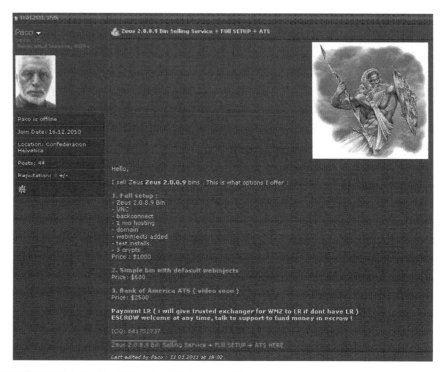

FIGURE 5.1 Ad for ZeuS Binaries in Underground Forum

- A month's worth of hosting.
- Back connection.
- A domain with Web injects.
- Three cryptographic packs.
- Test installs.

The second option is much less mature and feature rich. This retails for $600 and consists simply of the ZeuS 2.0.8.9 binary sold with default Web injection technology. The third offering this individual is marketing retails for $2,500 and consists of access to Bank of America Automated Transmission Systems (ATSs) available via the ZeuS Trojan or botnet.

This ad, though simple in nature, is effective as it demonstrates a then-current version of the ZeuS Trojan/botnet binaries available in a variety of configurations. It's important to note that this example was captured in November 2011, more than a year after the now infamous rivalry between the architects of the ZeuS Trojan and the SpyEye Trojan.[1] The significance of this ad lies not entirely in the availability of the binaries, Web injects, or ATSs, but rather, that it appeared approximately a year after the feud began (which, as Brian Krebs noted, is an uncommon affair; rare though not entirely without precedent[2]).

Furthermore, this example also illustrates that the alleged truce and surrender by the original author and steward of ZeuS, Slavik (a.k.a. Monstr), to the Spy-Eye crew (led by Harderman/Gribodemon) in March 2011 was a ruse. Most malware and botnet researchers believe the alleged surrender was nothing more than a false flag operation initiated to convince the SpyEye authors that their aggressive tactics (the demand for the surrender of the ZeuS source code by Slavik and agreement that he would not pursue another version of the botnet or perpetuate his struggle against them) had worked and that Slavik had in fact conceded defeat. The truth, evidenced by the ad shown in Figure 5.1, suggests otherwise, as we know Slavik released his code for a price, thus ensuring that ZeuS would continue on despite the efforts of a rival organization[3] (this version of ZeuS would also become the base for Ice IX[4]).

Figure 5.2 highlights a number of features that malware authors often advertise within the criminal underground. Some of the features that appeal to cybercriminals include the Bank of America Grabber (for user credential capture) and the CC grabber (useful to cybercriminals who either traffic in such commodities or retain them for their own use and illicit game). In Figure 5.3, the purveyor[5] of this variant of SpyEye is selling it for a remarkably low price ($150). The ad goes on to demonstrate the latest features associated with this version of the Trojan:

- Newest software features (associated with this variant):
    - Admin panel.
    - Formgrabber panel.
    - Gate installer.
    - Back connect.

FIGURE 5.2 SpyEye Version 1.3 Control Panel

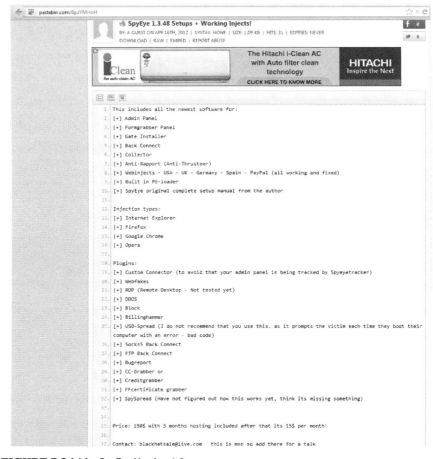

FIGURE 5.3 Ad for SpyEye Version 1.3

- Collector.
- Anti-Rapport (Anti-Thrusteer).
- Web injects (USA, UK, Germany, Spain, and PayPal).
- Built-in PE Loader.
- SpyEye original complete setup manual from the author.

- Injection types:
  - Internet Explorer.
  - Firefox.
  - Google Chrome.
  - Opera.

- Plug-ins:

  - Custom Connector.
  - Webfakes.

- RDP (Remote Desktop, untested).
- DDOS.
- Block.
- Billinghammer.
- USB-Spread.
- Socks5 Back Connect.
- FTP Back Connect.
- Bugreport.
- CC-Grabber or Creditgrabber.
- FFcertificate grabber.
- SpySpread.
- Three months of free hosting ($15/month thereafter).

The seller goes on to offer his e-mail address and encourages interested potential customers to contact him via MSN for a chat. In Figures 5.1 and 5.3, respectively, you can see that the sellers (suppliers) are keenly aware of what their target audience seeks in an illicit malware package such as the ZeuS Trojan or SpyEye. Figure 5.4 shows another (albeit simple) example of a seller's understanding of target audience and what drives their decision-making (purchasing) processes. This image was captured in October 2010. Notice the detail the seller provided and the language he used to describe this Trojan/botnet kit:

- Similar to ZeuS and in fact based on Version 2 of ZeuS.
- Core redesigned and enhanced to evade detection and threat mitigation technologies.
- Compatible with Internet Explorer and Firefox.
- Main functionality:
  - Key logging.

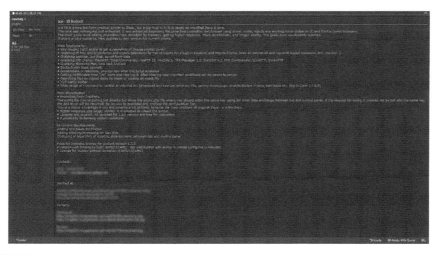

FIGURE 5.4 Underground Ad for the ICE IX Botnet

- Form grabbing.
- FTP client grabbing.
- Windows Mail, Live Mail, and Outlook grabbing.
- Socksv5 back connect.
- False certificates.
- Lateral movement capability.
- Support for diverse C2 on an infected host.

- Main advantages of the kit:
  - Tracker protection.
  - Higher response, longer vitality.
  - Updates and support.
  - Customization capabilities.

- Design goals:
  - Tracker evasion.
  - Higher response.
  - Increased stealth.
  - Longer vitality.

- In development:
  - HTTP fakes for Firefox.
  - Blocking/bypassing of SpyEye.
  - Dynamic algorithms for encryption.

- Pricing:

  - $600 for version that binds to the host (bot, bot builder).
  - $1,800 for builder license without limitations.

The ad in Figure 5.4 provides a degree of detail that goes above and beyond the earlier examples; however, it still represents a relatively simple yet effective message crafted to provide crisp examples of feature functionality, efficacy, and cost in addition to value added attributes such as upgrades and support. Examples such as these represent only the tip of the iceberg that exists within cybercriminal underground markets.

Figure 5.5 depicts a slightly more professional-looking front end for an online professional distributed denial of service (DDoS) attack product. The sellers operating this site stress their value proposition (their cost and efficacy) throughout the site. Note the attributes they have cited: their trustworthiness, their efficacy and expedience, the diversity of their offerings (from one hour to one month), and of course, their price point, which by many accounts makes this a rather inexpensive (though not the least expensive) DDoS product available.(see Figure 5.6)

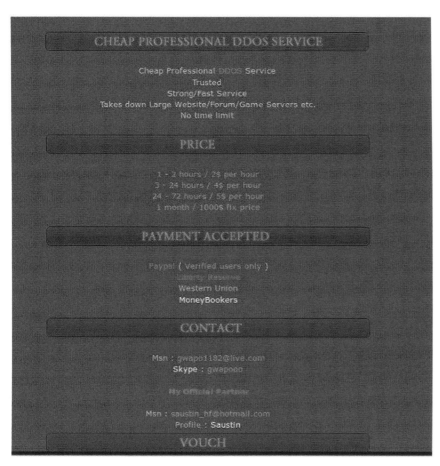

FIGURE 5.5 Example of an Illegal DDoS Service

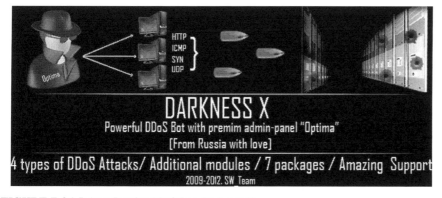

FIGURE 5.6 A Robust Russian DDoS Tool, Darkness X

SUMMARY

In this chapter we explored some common methods that cyber criminals use in identifying markets (existing or emerging) in which they seek to endeavor and profit. We discussed their understanding of the laws of economics, principally the law of supply and demand and their application of that insight in identifying and retaining their target audience's loyalty and patronage. We introduced concepts such as tools, tactics, techniques and procedures with examples of complex malicious code and content ecosystems that demonstrate the efficacy of TTTPs. This chapter provides an insightful look at the very tip of iceberg associated with criminal initiation and execution.

References

[1] www.sans.org/reading_room/whitepapers/malicious/clash-titans-zeus-spyeye_33393.

[2] http://krebsonsecurity.com/2010/04/spyeye-vs-zeus-rivalry/.

[3] www.symantec.com/connect/blogs/spyeye-bot-versus-zeus-bot.

[4] www.internetsecuritydb.com/2011/08/meet-ice-ix-son-of-zeus.html.

[5] http://blogs.mcafee.com/mcafee-labs/latest-spyeye-botnet-active-and-cheaper/attachment/1-4.

From Russia with Love

CONTENTS

INTRODUCTION

On June 12, 1987, the world listened as Ronald Reagan, the 40th President of the United States, addressed a crowd at the Brandenburg Gate near the Berlin Wall. Reagan was in Germany helping its citizens commemorate the 750th anniversary of the city of Berlin. It was the height of the Cold War, and the world was transfixed as its major political leaders, the United States and the Union of Soviet Socialist Republics (USSR), held the balance of world peace precariously within their grasp. It had been this way for the previous 42 years, since Japan had surrendered to the Allied Forces on August 15, 1945, marking the end of World War II. The United States, led by President Franklin D. Roosevelt; the United Kingdom, led by Winston Churchill; and the USSR, led by Joseph Stalin, had defeated the Axis Powers of Germany, Japan, and Italy, securing the delicate balance of power and freedom of the world once more. This would be a tenuous alliance at best, with the division of responsibility and control being vied for and culminating in the separation of Germany into two global blocs, East Germany and West Germany, and in the division of Berlin into four occupation zones: the French in the southwest, the British in the northwest, the United States in the south, and USSR in the east.

The next 42 years would see these former allies contend with one another in one of the greatest contests of will and unconventional warfare the world would ever see. Driven by espionage and intelligence-gathering efforts by a variety of organizations, two distinct camps developed, with the United States and its allies on one side and USSR and its allies on the other. This era in world history would see a fundamental shift occur in the economy of the industrialized world. The post-World War II global economy was burgeoning. The inception of what would become known as the Cold War[1] would see world powers shift their collective efforts away from their adversaries and toward themselves, thus ushering in a new era of intrigue and espionage. The respective economies of these world powers would blossom as the businesses of defense and offense approached their respective zenith points.

Then, on June 12, 1987, President Reagan delivered a speech that set in motion a chain of events that once again would propel these world powers toward a new era. In his speech, Reagan challenged Mikhail Gorbachev, the General Secretary of the USSR, to demolish the Berlin Wall as a sign of Gorbachev's commitment to increasing freedom in Eastern Bloc nations. This speech set in motion a chain of events that would see the end of the Cold War and the introduction and adoption of capitalism. This economic watershed would later prove to be an inflection point for cyber crime.

This chapter discusses how the events leading up to, and surrounding the Cold War relate to cybercrime today, and how the soviet mindset has influenced global cyber crime.

A BIT OF HISTORY

The speech Reagan delivered back in June 1987 would become known as one of the most famous and arguably most important speeches he ever gave. Reactions to the speech were varied. Upon reflecting on Reagan and his speech, Helmut Kohl, the former Chancellor of West Germany who was in attendance that day, said, "…[He] was [a] huge stroke of luck for Europe and the World."[2] Gorbachev stated in later years that if another leader was at the helm of the U.S. government at the time, he wasn't sure they could have arrived at the agreements they did during this era.[2] Though not all proponents of an East German communist government were moved by Reagan's speech, its impact was hard to ignore, as time and the actions of Gorbachev in 1989 and 1990 (with German reunification being imminent) would prove.[3]

Regardless of how world leaders and citizens felt, the storm of change that followed remains a subject of study to the present day. Revolutions took place in 1989 in Poland, Hungary, East Germany, Bulgaria, Czechoslovakia, and Romania, largely driven by civil disobedience and resistance guided by doctrines of nonviolence (Romania remained the only nation to overthrow its regime violently), with the fall of the Berlin Wall (as Reagan foresaw in 1987) representing a critical development in the world's understanding of communism and the separation between Eastern and Western nations. The net effect of these revolutions was the gross demise of communist-based nations in Europe. In 1991 the USSR dissolved, resulting in 14 countries declaring their independence, followed by similar revolutions in Albania and Yugoslavia. For many citizens of former Eastern Bloc nations, the world had effectively been turned upside down, creating no shortage of fear, uncertainty, and doubt as to what the future would hold for them and their families.

Initially it was rough going, with the standard of living dropping dramatically in most nations that had forsaken communism in lieu of a new system of government and economics. However, an unintended side effect that appeared in many newly re-formed nation states, such as Russia, was the birth of the new Russian business magnate (oligarchs) in paralleled and accentuated by disproportionate economic and social development and maturity. In his last work, *Collapse of an Empire*, Yegor Gaidar, who was Russia's acting prime minister from June 1992 to December 1992 in addition to being an instrumental figure in the metamorphosis of the Russian economy, stated that he believed the Soviet Union was in a prime position for abandoning communism. He felt (and as history has demonstrated, rightly so) that the former USSR had achieved a perfect storm of agricultural impotence that was unevenly coupled with the demand of its population for grain, which meant the USSR had to buy grain on the international market. At the same time, with the price of petroleum in flux during the latter portion of the 1980s, the USSR had to borrow funds to purchase the grain from Western banks. This set of conditions placed the USSR in a precarious position as they seriously impacted the USSR's ability to act on the international scene. Due to the decision to borrow money from Western banks to pay for grain to meet the demands of its most heavily populated areas, the USSR was severely impacted in its ability to send troops to address rebellions taking place in opposition of communism in Eastern Europe, largely because such action would likely result in Western banks refusing to grant the loan. Though many have since speculated about the conditions surrounding declining petroleum prices in the late 1980s that impacted the USSR's ability to purchase grain using foreign funds, the root cause remains dubious at best.[4] Such speculation includes whether the CIA, working with the leaders of Saudi Arabia, orchestrated a convoluted manipulation of the price per barrel of petroleum as a means to punish the USSR for its invasion of Afghanistan. Saudi Arabia did, in fact, increase its production of petroleum greatly, thus causing the price per barrel to drop dramatically on a global scale.

In the end, the allegedly progressive system of socialism based in communism was merely an imitation of feudalism. The dearth of personal freedoms of the common people along with the elite class's bent toward militarism embodied all the qualities supposedly abandoned in favor of progress. The common people, *the workers*, were relegated to the relative worth of the feudal serf; provided with the basic necessities to sustain life and not much beyond that. As a result, the conditions were ripe for change, socially as well as economically. For the everyman, the Soviet era proved to be an intricate time in which to live. There were plenty of opportunities within the Soviet system. Though communism espoused a system of equality unparalleled to any other, some members of the Communist Party could afford luxuries (real or perceived) that the law-abiding everyman could not. As a result, gaps existed during this era that many felt were insurmountable but that others felt could be overcome by circumventing the system through illegal and organized means.

Organized Crime in Russia

Organized crime in Russia has deep roots. It would be inaccurate to suggest that organized crime (like common crime) is pervasive and exists everywhere. In the case of Russia, organized crime's lineage and pedigree has existed in some form or another for more than 400 years. Yet in its modern form, we can focus on the years of 1917 through 1991, a period of Soviet authority that ensured the pervasiveness of organized crime in Russia and its neighbors.

This common lineage helps to explain the relationship between organized crime in Russia today, with its ties to the Russian political system, and its members. To understand this relationship, you must understand it in the context of Russian political and economic systems. In contrast to the well-known forms of organized crime in Italy, Japan, Colombia, and Mexico, Soviet or Russian organized crime was largely *not* based on ethnic or familial structures. In some societies, organized crime as a phenomenon has existed for hundreds or thousands of years. Russia is no different.[5] It is an accepted belief that in nations where urbanization is common and economic development and industrialization are the norm the demand for goods and services will be both stimulated and facilitated. Regardless of the nature, origin, or legality of the goods or services desired, the supply (availability) often dictates the rise in organized criminal activity. In times of prosperity, when there is a plentiful supply of unregulated goods, there is little to no incentive for organized crime entities to vie for entry into the market as suppliers. When the opposite conditions are present the incentive for organized crime activities increases to meet and fulfill unmet market demands. These demands could be for a variety of illicit things: drugs, firearms, or pornography, or even unauthorized access to compromised systems that can be used to promote myriad goals and initiatives. The nature and number of illicit opportunities available in a particular geographic region also influences the presence of organized crime entities along with the degree to which such exploitation can be achieved. Unlike organized criminals being borne through waves of immigration into the United States, seeking to elevate themselves through illicit means, sharing a common ethnicity, culture, and language, Russian organized criminals (with the plausible exception being those in certain areas of Armenia, Chechnya, Georgia, and Uzbekistan) sought to achieve their goals of illicit gain by collaborating with other like-minded individuals or groups regardless of their ethnic makeup. The bonds that tied in the case of these organized professional criminal syndicates were *economic* as opposed to *ethnic* or *familial*.

THE COLD WAR AND ITS SIGNIFICANCE TO CYBERCRIME

As we have discussed, the end of the Cold War ushered in a great deal of change in the former republics of the USSR. Organized crime began to flourish, with criminals who were ready, willing, and able to conspire with one another to accomplish their illegal acts. In order to achieve this, a degree of trust was necessary among these criminals. Trust among coconspirators, or the ability of conspirators to trust one another in all matters (when dealing with the authorities, dealing with money honestly, and so on) is imperative. With respect to organized crime organizations in Russia and Eastern Europe, a shared ethnicity (complete with a common language, background, and culture) is not essential, nor is it the norm. This is due largely to how the professional criminal class developed within the prison systems of the former Soviet republics, from the Stalinist period of approximately 1924 to the modern era.

Criminals within the prison system adopted the behaviors, rules, values, and authorizations necessary to bind themselves together in what has been commonly referred to as the "thieves world,"[6] led by the elite "Vory v Zakone."[7] The Vory v Zakone (also known as the "Thieves in Law") created and maintained bonds and an atmosphere necessary for achieving their goals. During the Soviet era their activity consisted of operating illegal enterprises with connections to legitimate and black market entities. These operations continued to flourish in the post-Soviet era as well, especially operations based on the "Nomenklatura" system.[8] This system saw members of the Russian government bureaucracy forge connections with members of the thieves' world. As a result, evidence of the first modern examples of Russian organized crime began to come to light in the late 1960s. At the top of this criminal infrastructure were high-ranking members of the Russian government and bureaucrats. The second tier of this model consisted of underground or shadow participants willing to exploit themselves, their jobs, or their connections for profit. Their participation in "Shadow Economies" was strictly "off the books" and saw the payment for such goods and services flow directly to members of the second tier. At the bottom were the professional criminals who operated various illegal activities, such as drug trafficking, gambling, prostitution, and extortion. It was these lower-level tertiary players who, in working with members of the first and second tiers, set forth the model that supports the natural entry of such criminals into crimes leveraging the Internet.

And why not? The risk of this has historically been quite low in comparison with the reward. It was the perfect storm for professional criminals well versed in earning from less lucrative, higher risk activities. Upon hearing Reagan's call, the former USSR, along with all those nations in Eastern Europe that subscribed to a particular brand of Marxist-influenced socialism, would be forever changed, as would the criminal enterprises which for the previous 72 years stood in the shadows.

A Look Back at the Birth of Economic Prosperity and Chaos

Before we continue, we should take a step back and analyze how economic prosperity, and the chaos that ensued, have played a role in cybercrime in Russia and Eastern Europe.

Vladimir Lenin was a prominent figure in the Russian Revolution of 1917. He was not only one of the leading political figures of his day, but also one of the most noteworthy revolutionary thinkers of the 20th century. His role in planning the Bolshevik Revolution beginning in October 1917, in addition to his leadership and vision, aided in its success some three years after it began.

A principal element of Lenin's philosophy for a post-tsarist Russia was predicated on a seemingly simple edict: Peace, Bread, and Land, a concept that became a slogan of the October revolution. Lenin (who would assume the first true leadership of the newly formed USSR) knew the power and significance of these words to the average Russian. *Peace* would be predicated on Russia's removing itself from its involvement (based on treaties and alliances) in the costly efforts associated with World War I. *Bread* represented the need to change and improve the day-to-day living conditions of the average Russian, which up to that point was dependent upon a number of factors, not the least of which was the benevolence or lack thereof of the land owner himself. *Land* represented what Lenin believed to be a principally important concept for every man, woman, and child in Russia: the opportunity to own land communally as opposed to being anchored to it via the archaic system of feudalism and serfdom present in Russia at the time of the revolution.

It was a post-communist world where capitalism and open markets, legitimate and illegitimate, were ushered into the everyman's reality with a boldness not seen since the revolution. Businesses began to thrive and international trade opportunities began to flourish. To everyone who witnessed this transition, it seemed as though capitalism suited the people of the former USSR quite well. In fact, it could be argued that due to the expeditious nature of the transition from communism to capitalism, the population's level of comfort (reflected by its embrace of new business ventures legitimate and illegitimate alike) with the evolution they were experiencing was commendable.

In fact, it was noteworthy. Yet, as we have pointed out, it was not without its warts. Russia and Eurasian nations carved out of the former USSR have long been (and remain today) the arguable leaders of malicious cyber activity and cybercrime, though some would speculate that the United States and China are close seconds in this race. As we have discussed, Russia's post-communist adoption of capitalism and the subsequent adoption of capitalism by former members of the USSR have led to the demonstration of geographically unique challenges coupled with socioeconomic ones. These challenges, when assessed alongside what can only be characterized as an often punctilious political system, led to the creation of a perfect storm condition in which criminal activity, including the budding Internet-based cyber variety, would bloom and flourish.

Yet even within this maelstrom, another vector emerged that, when considered in the context of the other variables discussed earlier in this chapter, warranted further examination and consideration. Russia and many other members of the former USSR had for many years encouraged the production of tens of thousands of gifted minds (mathematicians, engineers, scientists, and others) seeking to enter a job market that simply could not accommodate and sustain candidates of their ilk and ability. This lack of opportunity, coupled with an indifferent and accepting attitude toward corruption by younger Russians and Eurasians, would lead many into the service of both amateur criminal organizations and their more sophisticated peers, the Russian Mafia.

At this point, you may be wondering why this would manifest and become so arguably attractive to so many otherwise promising and talented individuals. For many it is the allure of the taboo. For others it is the prestige and money earned by violating vulnerable individuals and organizations the world over. For others still it is the idea of earning vast amounts of money for more serious criminal entities, thus garnering the societal and cultural rewards associated with climbing the ladder of professional criminal activity. The political leaders of the regions in question, especially those of Russia, often (due to their affiliations in primary and ancillary criminal hierarchies) fail to be of much use in discouraging these problems and the behaviors associated with them. Researchers have often noted that this level of indifference is normal unless a

Russian organization is targeted and harmed. As a result, the Russian cyber-crime underground has evolved into a sophisticated, if loose-knit, community with its own periodical literature and cultural mores. The "Russian hacker" has become a being of allure and intrigue. He is a fictional character while also being the real-world bane of security analysts and law enforcement agents around the world. He is a stereotype; a model that some emulate and others fight. Russia and the former states that once comprised the USSR do have a large population of talented hackers that are under less pressure from the law than their counterparts elsewhere[9] in the world to date. Though this is changing, as we have seen in cases associated with the takedown of various parts of the ZeuS banking Trojan/botnet gang and the DNS Changerbot gang, among others, it is imperative that firms doing business in Russia and her neighbors not only be able to secure themselves from the relentless challenges of cyberspace, but also consider other, often more difficult problems.

WHY THE RUSSIANS UNDERSTAND ECONOMICS BETTER THAN YOU DO

As we have seen, Russian and Eurasian post-communist capitalism has enabled people to embrace and thrive in a neo-capitalistic society. It is a society that embraces a cultural tolerance of corruption in government. It is also a society that observes (more often than not) a lax approach to the investigation and prosecution of crimes occurring within its national boundaries unless the firms or organizations being targeted are Russian or Eurasian.

Therefore, it is important to recognize how base economic principles in addition to nontraditional socioeconomic realities (generally tied to black market activity or illicit markets supported by corrupt officials) impact the daily lives of the average citizens of these nations. The trends associated with Russian and Eurasian cybercriminal activity depict this and show unequivocally the dedication to profit from their illicit actions.

The Russian Carder Scene

One area that illustrates this well is the Russian carder scene. It is arguably the most heavily populated in the world along with being the most active (defined by the number and frequency of monetary transactions) of any in the world with the plausible exception being those of the United States (though there have been mass migrations from U.S.-based carder sites to Russian sites due to several U.S.-based Department of Justice led operations such as 2004's Operation Firewall and 2006's Operation Cardkeeper). One area that demonstrates Russians' understanding of tapping into a potential addressable market comes in the form of their recognition of the value of specialization. Many Russian carders, for example, specialize in the development of unique and sophisticated attack tools along with service-based models made available to interested third parties seeking to capitalize through the trafficking of stolen credit, debit, banking, and other financial account data. This data has a time to live (TTL) within the underground, and depending on what is being sold or marketed, the value diminishes exponentially the longer it is in hand.

So, What's for Sale?

The data being harvested and sold within these illicit underground forums (carder-centric and non-carder-centric alike) can be identified and classified in many categories beyond those usually associated with these environments. As exemplified in Figure 6.1, marketplaces like the Omerta[1] Russian Card Readers Forum provide a place where "verified people… [can engage in] serious deals". In the more sophisticated environments, especially those located in Eastern Europe and parts of Asia Pacific, this may include (but is not limited to) primary research (intellectual property), countermeasure research (law enforcement, professional security analysts, intelligence firms), and, more recently, confidential documents related to national security. These forums and the data contained within, along with the conversation, guidance, and more than occasional trolling, enabled would-be malicious attackers to sharpen their skills and methodology. This "education" could also include guidance on target/target environment selection as well as human intelligence concepts. It could be argued that those operating these forums (in addition to frequenting them) believe and endorse the idea that increasing the likelihood of their success requires studying their adversaries (in this case, law enforcement and information security professionals).

This mindset is loosely reminiscent of military reconnaissance or intelligence operations. Through developing, proving, and trafficking this data, attackers

FIGURE 6.1 Omerta.cc Russian Carders Forum

---

[1] "Omerta" refers to the code of silence respected by the Sicilian Mafia.

increase their chances of pulling off stealthy attacks while conducting campaigns of disinformation (false flagging is a distinct possibility in these cases) and obfuscation in order to throw off law enforcement and information security personnel. So, the answer to the question "What's for Sale?" in the various forums could be everything and nothing at all, depending on what forums are being observed, the nationality and primary language(s) of the parties involved in the dialogues, how ambitious someone is, whether someone is "known" and trusted, and how much someone is willing to pay and what the seller is asking for his wares.

CYBER THIEVES

The former USSR has become culturally, educationally, and legislatively ripe for the type of criminal activity that we are most genuinely interested in and that you will likely be familiar with. However, it is important to note that just as with any professional, syndicated, organized criminal effort, the potential for convergence is both real and likely. In the case of cybercrime being sourced, supplied, and in many cases initiated from within former republics of the old USSR, this is and remains the case today.

Some of you may be wondering how it could possibly not have evolved in this direction given how the communist regime was dismantled with the USSR and other former Eastern Bloc (Eurasian) nation states that shared the same philosophical and economic views for the better part of a century. In the course of researching and writing this book, we came to share a common set of beliefs with respect to the data collected and the panoramic view it provided into the early stages of the conditions associated with and leading to the evolution of this new era in post-Soviet criminal activity:

- Speculation is a futile exercise; facts speak volumes as do corresponding/supporting data points.
- The investments made by the former government of the USSR in advanced mathematical, scientific, and computer science studies in the pre- and post-university system, military complex, and intelligence communities were substantial. They led to a considerably high number of well-educated, well-trained people who found that their ability to earn a living (commensurate with their level of effort in honing their educational skills and craft) was dramatically less than what they could have hoped for. This lack of hope is a universal condition associated with the entry of both amateurs and professionals in this vein of criminal activity.
- Legal loopholes and a fundamentally uninterested (historically, though this has begun to change over the past several years, as noted in Joseph

Menn's outstanding piece *Fatal System Error*)[10] and not entirely legal law enforcement system made it virtually impossible to prosecute criminals for such activity, regardless of what type of illicit actions were being initiated and profited from.

- The low cost of operations in the former USSR republics, coupled with a virtually inexhaustible sea of highly skilled information technologists and security personnel, are viewed as key factors in both the growth and success of criminal elements in this part of the world, among them the Russian Business Network.[11]

- The odds were in favor of Russian and Eurasian cybercriminals as opposed to average IT operations and IT security professionals. We'll explore this idea in more depth in the next section, though it is important to consider here as well. Does this mean Russian cybercriminals are more intelligent than their white-collar counterparts working in enterprise security? No, not necessarily. Although the context of the statement can be easily misunderstood at first glance, one point is quite clear: Those who opt for a certain level of reward in most cases accept the risk associated with gaining (potentially) said reward. Russian and Eurasian cybercriminals are well aware of the rewards and risks (in many cases, performing formal cost benefit analyses that would rival those conducted by legitimate commercial enterprises), as well as the realities associated with the possibility of being caught. The net effect of this, coupled with their level of determination, willingness to diversify their business models to capitalize as quickly and grossly as possible, ample funding, and comprehension of the limitations enterprises and end users struggle with (among them governance, budgets, knowledge, and education), places them in a position that looks dominant. This position is not infallible or impregnable. It simply requires deep consideration and serious thought when analyzing these groups and their operations.

CYBER RACKETS: BOTNETS, MALWARE, PHISHING, AND THE RISE OF THE RUSSIAN CYBERCRIMINAL UNDERGROUND

The Internet has made certain things complicated, especially things of a jurisdictional and legal nature, on a global scale more so than anyone could have ever foreseen. What is considered "illicit" or "illegal" to one person, party, or nation state may or may not be so for other people, parties, or nation states. As a result, the ensuing deluge of criminal activity originating in geographically disparate regions impacting targets/victims in other regions has reached pandemic heights.

Nowhere is this truer than in the case of Russian and Eurasian (Eastern European) cybercriminals. Cultural views and perceptions of criminal behavior aside, from a law enforcement perspective this was a largely unfamiliar phenomenon. There was precedence for dealing with acts of unlawfulness in one's given nation and municipality. There was protocol, procedure, and process (in the United States there was and is due process,[12] though this is not a universal absolute in all nations) that law enforcement agents and officials could execute within the legal boundaries of their respective legislative bodies and doctrine in the event one sought to prosecute an individual or group of individuals involved in "illicit" activity within the jurisdictional boundaries that a given municipality, state, or nation was responsible for. But what if the offending party initiated and completed one or several acts that violated the law within a municipality, state, or nation from halfway around the world? Was there a precedent governing that? Complicating matters further, what if the nation in which the offending party was operating was considered a "nonextradition" nation? And moreover, what if one could not prove unequivocally via attribution that a given person or persons were responsible for a given act or acts of an illegal nature?

Unlike with traditional criminal acts, the physical proof was weak, if present at all. Therefore, more advanced methods (many of which were borrowed directly from the discipline of criminal forensic science) would become necessary, if not mandatory, in any and all serious investigations, reported to the authorities or investigated privately so as to manage publicity. Professional criminals who became aware of the magnitude for gain that the Internet represented understood this, and as a result, they took appropriate steps to capitalize from them.

The Rise of Hacking

The year 1994 would be significant for Russian and Eurasian hacking and organized cybercrime operations. On March 5, 1995, Vladimir Levin,[13] a Russian citizen, was arrested in the transfer lounge at Stansted Airport in London, by members of Scotland Yard at the behest of the U.S. government with whom the United Kingdom (Great Britain) has an extradition treaty, for his role in the illegal access and compromise of several large corporate customers of Citibank N.A. Levin and his crew allegedly accessed the account information of these customers via insecure dial-up wire transfer services, transferring funds to accounts set up in multiple nations. The Levin crew attempted approximately 40 transfers to accounts set up in Finland, Israel, the United States, Germany, and the Netherlands.[14] When Levin was arrested he was awaiting an interconnecting flight from Moscow. In addition, three members of his crew were arrested (one member in Tel Aviv, one in Rotterdam, and one in San Francisco, while attempting to withdraw funds from the shell accounts). Levin was eventually extradited to the United States, though his attorneys in the United Kingdom fought against it (an appeal was rejected by the House of Lords in 1997).

Levin was eventually convicted after being remanded to the custody of the United States in September 1997. He eventually pleaded guilty to one count of criminal conspiracy to defraud and to stealing $3.7 million. He was sentenced in 1998 to three years in jail and ordered to pay restitution of $240,015. In the aftermath, Citibank claimed that all but approximately $400,000 was recovered of the $10.7 million stolen.[15] In 2005 an alleged member of Levin's St. Petersburg hacker crew (claiming to be one of the original Citibank N.A. penetrators) published a memorandum under the name of ArkanoiD on the Russian Web site provider.net.ru. In his memorandum, AranoiD asserts that Levin was merely a system administrator who had acquired information on how to penetrate the Citibank N.A. systems in order to exploit them for $100.

Regardless of Levin's acumen or facility with hacking, he and his crew managed to pull off the first well-publicized example of Russian cybercriminal activity. Their efforts would set a tone for the next two decades that would be felt the world over. ArkanoiD would go on to assert that Levin and company would exploit the Citibank N.A. systems via X.25 system exposures versus Internet-based connectivity. But the transport mechanism was irrelevant in light of the end game: the compromise of Citibank for nearly $11 million.[16]

Enter the Bots

Botnets have a long and checkered history in modern computing. First discussed in serious computer science and academic communities, botnets represented the extension and epitome of truly decentralized computational power that could be harnessed for a common purpose. Initially this purpose was not malicious, but soon it was realized that botnets could be established devoid of intent and leveraged by the botmaster to achieve his or her ends.[17]

At its most base level, botnets are nothing more than tools,[18] and there are as many motives for using them as there are variants of them. The most common (and arguably, most potent) use is to conduct some sort of cybercrime. This goes well beyond the predictable tool of ignorance, the distributed denial of service (DDoS), which, in light of recent events, has proven to be more formidable than many had previously considered, and is arguably at the epicenter of some of the most advanced cybercriminal operations the world has seen. Botmasters (those who architect, harvest, and manage botnets) are involved in many operations related to Crimeware as a Service (CaaS). Notable activities associated with botnets (especially of the Russian and Eurasian varieties) are as follows:[18]

- DDoS attacks.
- Spamming.

- Traffic capturing.
- Key logging and credentialed account theft.
- Malware distribution and promulgation.
- The installation of advertisement addons and browser helper objects.
- Google Adsense abuse.
- The compromise, exploit, and attack of IRC and Instant Messaging chat networks.
- The compromise, exploit, and manipulation of online gaming environments.
- Gross identity theft.
- Extortion.

DDoS attacks are staples of activities associated with botnets regardless of their level of sophistication. In fact, they work just as well today as they did a decade or more ago and have been proving to be useful tools in geographies not previously believed to be traditionally afflicted by them. During 2010 and 2011 this was proven to be the case through operations launched by the Internet Hacker Collective (IHC) Anonymous. Their use of the DDoS attack proved effective and crippling to businesses and governments the world over in support of their individual and collective messages. It is unclear today as to whether members of the Russian and Eurasian cybercriminal underground were involved in these operations, though it cannot be ruled out entirely. Spamming, like DDoS, was a problem thought to be localized in time and space long since past in terms of significant efficacy.

Though there is data that suggests there were lulls in activity along with shifts in direction from SMTP-based spammed activity to HTTP over the past five to seven years, spam has reemerged and is becoming noteworthy once more. There are several reasons for this, not the least of which is a rise in the occurrence and effectiveness of spear phishing and targeted spear phishing attacks against individuals and organizations.

Botnets are the perfect promulgation tool for these activities. In 2007, the Storm botnet, believed to be owned and operated by the Russian Business Network, came on the scene and quickly rose to prominence promulgating malicious spam attacks across the Internet. It was a superbly crafted piece of code observed DDoS'ing[19] anyone seeking to attack its command and control servers. At its height, it accounted for roughly 2.6 million zombie bots in its network.[18] The Storm botnet declined in activity and prominence, moving toward block obsolescence, in 2008. A great deal of speculation surfaced as to what led to the decline. Some argued it was due to a lack of interest on the part of the architects in maintaining the botnet, while others argued that it was more fundamentally related to the release of tools that allowed for successive hijacking of the botnet.[19]

In 2010, McAfee confirmed rumors related to the existence and anticipated release of the next Storm botnet (code named "Dark and Stormy").[20] Russians and Eurasian cybercriminals have, for many years, favored malicious code and content attacks such as command and control-enabled Trojans for a variety of reasons, not the least of which is their elasticity and adaptability to environmental shifts and changes. When discussing matters regarding the profitability of effort and operation related to botnets, there is no greater power to have in one's repertoire than adaptability coupled with efficacy.

A great example is the ZeuS Trojan (also known as the ZeuS botnet). ZeuS is a complex beast. It was first identified in July 2007 and it is as diverse as its wielders. ZeuS enables (among other things) the theft of credentialed account information and form grabbing. Historically, ZeuS was most likely to be contracted via surfing (or being directed to) an infected Web site or via a phishing scam. In 2009, one variant and instantiation of ZeuS was discovered by the research team at information security firm Prevx that yielded a cache of more than 74,000 FTP credentials to accounts including those belonging to NASA, Monster, ABC, Oracle, Cisco Systems, Amazon, and BusinessWeek, among others.[21] The monies stolen in this operation were smuggled from the victim accounts through unauthorized transfers of thousands of dollars at a time, and often were routed through an intricate web of fraudulent accounts controlled by the thieves via money mules. The money mules in the United States were recruited from overseas and encouraged to create fraudulent bank accounts using false identities. Once the money had reached the mules' accounts, the mules would wire it to their handlers in Eastern Europe or attempt to smuggle the money out of the United States (a much riskier proposition). All told, approximately $3 million was stolen. It is believed that this gang alone had, over time, attempted to steal more than $220 million while successfully managing to abscond with more than $70 million.

SUMMARY

This chapter has provided an in-depth historical perspective of how the Cold War influenced and created a large part of what is a critical player in the cyber-criminal reality. The events that led up to it forged the culture that permeates even today in the cyber criminal world.

References

[1]   www.worldwar2history.info/war/causes/Cold-War.html.

[2]   www.usatoday.com/news/world/2004-06-07-reagan-world_x.htm.

[3]   www.usatoday.com/news/wo.

[4] www.foreignpolicy.com/articles/2011/06/20/everything_you_think_you_know_about_the_collapse_of_the_soviet_union_is_wrong.

[5] www.ncjrs.gov/pdffiles1/nij/187085.pdf.

[6] This is empty reference

[7] www.pbs.org/wgbh/pages/frontline/shows/hockey/etc/glossary.html.

[8] www.jstor.org/discover/10.2307/152994?uid=3739656&uid=2129&uid=2&uid=70&uid=4&uid=3739256&sid=55989298143.

[9] Jellenc E, Zenz K. Global Threat Research Report: Russia. An iDefense Security Report by the VeriSign iDefense Intelligence Team; 2007.

[10] New York: PublicAffairs, 2010. Fserror.com.ISBN: 1586487485.

[11] www.bizeul.org/files/RBN_study.pdf.

[12] www.law.cornell.edu/wex/due_process.

[13] www.publications.parliament.uk/pa/ld199798/ldjudgmt/jd970619/levin.htm.

[14] media.ais.ucla.edu/BTseminars/fbi_slides.pdf.

[15] www.cab.org.in/Lists/Knowledge%20Bank/Attachments/64/InternetFraud-VL.pdf.

[16] www.thekomisarscoop.com/2002/08/us-investigators-missed-russian-mob-in-ny-bank-scandal/.

[17] www.nanog.org/meetings/nanog32/presentations/kristoff.pdf.

[18] www.honeynet.org/node/52.

[19] blog.washingtonpost.com/securityfix/2007/10/the_storm_worm_maelstrom_or_te.html?nav=rss_blog.

[20] blogs.mcafee.com/mcafee-labs/dark-and-stormy-comeback-of-a-botnet.

[21] www.thetechherald.com/articles/ZBot-data-dump-discovered-with-over-74-000-FTP-credentials/6514/.

CHAPTER 7

The China Factor

INFORMATION IN THIS CHAPTER:

- Economic Growth
- Industrial Espionage
- Reducing the Risk of Industrial Espionage
- Security Spend Model

CONTENTS

INTRODUCTION

The essence of this chapter can be summed up by a well-known quote from the movie *Iron man. Tony Stark says the following.* "My old man had a philosophy: Peace means having a bigger stick than the other guy." The bigger stick that China carries combines the factors of economics, geopolitics, and mass population. All of these play a critical role in China's ability to successfully run, against any target it wants to control, a variety of cyber operations that will lead to the theft of intellectual property. Furthermore, the alleged non-state-sponsored groups within China that are conducting these acts are huge in terms of number of members. With this in mind, this chapter covers the drivers that continue to make China one of the biggest cyber security threats on the Internet.

ECONOMIC GROWTH

China is the largest country in Asia, and its land mass makes it the fourth largest country in the world, with a population of approximately 1.3 billion people[1]. You might think that the large population of China, when compared with the smaller population of the United States of roughly 320 million, means that China's gross domestic product (GDP) is much larger than the U.S. GDP. However, Figure 7.1 shows that this is not the case.

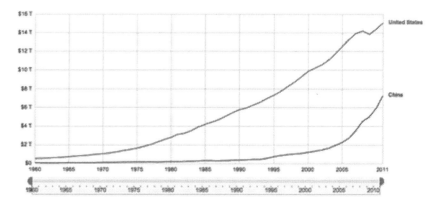

FIGURE 7.1 GDP of the United States Compared to That of China

When you look at the formula for determining a country's GDP:

$$GDP = C + G + I + NX\,^2$$

where:

- "**C** represents all private consumption, or consumer spending.
- **G** is the sum of government spending.
- **I** is the sum of the country's businesses spending on capital.
- **NX** is the nation's total net exports, calculated as total exports minus total imports (NX = Exports – Imports)" [2].

In reviewing this formula and looking at the trend line from China vs the United States, we can see a rapid spike in their GDP starting 2005. This rapid spike takes in to account many variables as mentioned in the formula but does not take into account some of the technology advancements that they have made in the last 6 years regarding their dominance in above ground rail systems and other technologies that are discussed below.

Economic growth is a key factor that enables advancements that measure a country's success and power. And in that regard, China is considered to be experiencing a new industrial revolution. For example, in the mid-2000s, China had no mass rail transit systems. "Today, it has more than Europe and by 2012, it will have more than the rest of the world put together," [3] according to an article by the BBC. You might not think this is significant, but many countries have spent decades building their mass transit systems, while China achieved its goal in less than a decade.

What is especially interesting about rail transit in China, as reported by the BBC, is that under a new proposal, known as "indigenous innovation," foreign companies bidding for Chinese government contracts will have to share the existing know-how. They will also be required to conduct any new research and development work in China. For some companies this will prove to be an unacceptable condition, according to Brenda Foster, head of the American Chamber of Commerce in Shanghai. "It will keep American companies from being able to compete in the Chinese domestic market," she says. "For some companies, that could actually put them out of business." [3] This is very important to point out, because China has the skill and cyber power to gain vast amounts of technological knowledge from other companies around the world, enabling China to advance in certain areas faster (and sometimes better) than other countries. China's indigenous innovation concept is beneficial because China has the capital to fund many projects that are attractive to new and established companies that are willing to share or give away their technology if doing so will result in major revenues. China can then capitalize on this technology to enable its own expansion and growth.

Although this activity might scare you if you do not live in China, you must realize that for any amount of money or key industrial contract within China they will award a lot of money for the right contract. China provides large profits for foreign companies, thus increasing the status of these companies on Wall Street and might increase the propensity to move operations and bid on contracts within China. Consider what happens when the United States outsources to foreign countries. Conducting technological business in the United States is very expensive. As a result, many U.S. companies decide to outsource engineering and manufacturing to foreign countries because doing so decreases capital investments and increases profit margins. For example, the salary of a typical full-time engineer in the United States is approximately $150,000; in India and China it is approximately $15,000. Offloading this cost increases the profitability of any corporation doing business with China, but the increase in profits only provides a competitive advantage to China.

INDUSTRIAL ESPIONAGE

The Chinese are very clever in terms of playing both sides of the fence. Indigenous innovation nets China legitimate access to technology that advances its own technology and increases its economic standing. [4] What's important to note is that China is spending $106 billion on Defense, which is 1.28 % of its GDP.

Indigenous innovation also nets China illegitimate access to technology. This access is accomplished through industrial espionage. There are several use cases around industrial espionage that involve China stealing information from U.S. corporations. The reason why any foreign entity goes after another country's technology is that it doesn't want to spend the time, resources, and capital investment that are necessary for acquiring new technology. According to an article available on The Economic Populist Web site, the following are a few industry verticals that China is actively targeting: [5]

- Communications companies: 23%.
- Aerospace and defense: 18%.
- Computer hardware and software: 14%.
- Energy or oil and gas: 10%.
- Electronics: 10%.
- Other, of which the financial sector was the largest component: 25%.

Cases of Industrial Espionage

To put this in perspective, in this section we'll take a look at some recent cases of industrial espionage. These are important because they demonstrate the damage that is caused when an organization's security is breached.

The first case we'll discuss, known as the Byzantine Foothold attack, [6] was allegedly tied to China and targeted many companies worldwide, including Xerox, Volkswagen AG, Yahoo!, and Hewlett-Packard, to name a few. All of these companies had specific command and control servers that were sending data outside their networks to similar addresses in China. According to an article on Bloomberg.com, Xerox performs back office and human resources support for many companies. Although the cyber actors were going after Xerox, they probably received more than they bargained for because they may have gained access to many of the companies that Xerox performs back office support for.

Meanwhile, according to a recent article in *Business Week*, [7] China is targeting the biotechnology, telecommunications, nanotechnology, and clean energy industry vertical markets. Toward that end, one of the examples cited in the article concerns a U.S. metallurgical company that lost, to hackers in China, technology that cost the company $1 billion and 20 years to develop.[5] This case is significant because the perpetrator of the attack was reportedly a non-state-sponsored entity. Information operations of this scale require a lot of time and persistence because the information that is stolen can take months or years to steal.

In a recently written testimonial to the U.S. government, Richard Bejtlich, chief security officer at the information security company Mandiant, stated some surprising statistics that all authors and other experts within the cyber security are not surprising facts in most breaches: [8]

- "94% of victims learn of compromises through third parties, while only 6% discover intrusions independently. Victim organizations do not possess the tools, processes, staff, or mindset necessary to detect and respond to advanced intruders.
- The median number of days that elapse between the compromise of a victim organization and detection or Mandiant involvement is 416 days. Incredibly, this number is an improvement over past intruder 'dwell time' measurements of 2–3 years.
- Advanced intruders installed malware on 54% of systems compromised during an incident. They did not use malware to access the other 46% of systems compromised during an incident, meaning relying on tools that find malicious software misses about half of all victim computers.
- Mandiant observed intruders using stolen credentials in 100% of the cases it worked in 2011. Intruders seek to use legitimate credentials and access methods as soon as possible, because they can then 'blend in' with normal user activity."

These facts are somewhat sobering and should be a wake-up call for any organization with an Internet connection. Most breaches are typically found by a third party, and it's likely that the third party is law enforcement (local or federal) and that the activity has been going on for at least a year. The amount of data that can be taken in 60 seconds with a 1Gbps connection can exceed approximately 5GB of data. Imagine the amount that can be taken in 416 days. According to Bejtlich, this number used to be higher, so when you read in the media that certain breaches have been ongoing for several years, it shouldn't be surprising.

Tier 1 and Tier 2 Security

This brings us to an important point regarding the technologies that are supposed to protect our infrastructures from breaches and cyber attacks: The cost of tier 1 and tier 2 security products can be very high. As a result, most organizations today typically deploy only tier 1 security technologies and fail to invest in tier 2 technologies as well. This is why a company whose security has been breached can take up to 416 days to discover an advanced or targeted attack, and why the breach is usually not even discovered by the company's internal security team.

Let's take a look at what constitutes tier 1 and tier 2 security technologies so that we can put this into perspective in terms of China and its cyber capabilities today.

Tier 1 Security Technologies

In large corporations, tier 1 security technologies are the basic tools for building out what is considered best security practices or defense in depth. According to today's security manuals and compliance regulations, the following are considered necessary for building out a reasonably secure infrastructure:

- Firewall or next-generation firewall.
- Desktop anti-virus tool.
- Secure Web gateways.
- Messaging security.
- Intrusion detection/prevention systems.
- Encryption (in transit or at rest).
- Security information event management.

At this point you may be wondering: If these technologies are so good at providing protection, why are we seeing such a high number of serious security breaches? The answer is because tier 1 security technologies are only good for attacks that are known by the security community, and to be fair, some vendors claim zero-day protection for vulnerabilities that are not known to the general public.

In the case of China, we are dealing with a country that has a proven track record of industrial espionage that has cost corporations billions of dollars in time and research. It takes more than a year for an organization to realize it has been compromised because the Chinese are clever enough not to use well-known attacks. Additionally, China has several hacker organizations to contract targets of interest, or it has its own government cyber capabilities that are probably the best in the world. The Chinese are also smart enough to realize that most organizations are going to be deploying tier 1 security technologies and possess the capability to get around these technologies. However, it's very important to note that tier 1 security technologies are not useless. They are still needed to keep out the average hacker, but they will not stop the sophisticated hacker. This brings us to tier 2 technologies.

Tier 2 Security Technologies

Tier 2 security technologies are often purchased after a major breach occurs. They are used by some of the most sophisticated organizations in the world that understand they have to combine tier 1 and tier 2 security technologies to provide the security that is necessary for reducing their risk profile. These tools include the following:

- Network forensics.
- Desktop forensics.
- Data leakage protection (network/desktop).
- Behavior-based analysis.
- Security intelligence feeds.

In a recent interview, the head of the National Security Agency (NSA) was quoted as saying the security technology available today only protects approximately 80 % of the attacks on the Internet. Tier 2 security technologies are needed to fill the 20 % gap because they go beyond traditional pattern matching and signatures for known attacks. Tier 2 security technologies have the capability of identifying abnormal behavior in transit and on the host. Additionally, some of these technologies have the capability to model the behavior of a given file or binary that might be considered dangerous for the receiving host. For example, consider a user who downloads what he or she believes is a normal PDF file on a specific topic of interest. The downloaded file contains embedded malicious JavaScript code and, when the person opens the file, the JavaScript executes and the person is now "owned." Most tier 1 security technologies would never be able to stop this type of attack because they are mostly focused on named threats. Because of this, the security industry and regulatory compliance organizations must include tier 2 security technologies, along with tier 1 technologies, in their security best practices.

REDUCING THE RISK OF INDUSTRIAL ESPIONAGE

There is no magic formula for protecting your critical intellectual property from China, but there are steps you can take to make sure you reduce your chances of getting a call from a third-party company, notifying you that you have been breached. Earlier in this chapter we mentioned a company that lost more than $1 billion in research and time. It's likely that the company only had tier 1 security technologies in place. The typical IT spend on tier 1 security products is approximately 3 % of a company's overall IT budget. With that said, most organizations will use those security dollars to achieve or maintain regulatory compliance, and often to check the box on what are considered security best practices. Tier 2 security technologies are often viewed as a "nice to have" benefit, or they are purchased after a serious breach occurs. The broader security industry must adopt and place as much emphasis on tier 2 security technologies as they do on tier 1 security technologies. Realistically, nothing will secure the greatest security risk of all, and that's the front door. Additionally, there is no patch for stupidity, which is social engineering by target or accident. What this is really conveying is that anyone who wants information will go beyond just the pure technical means to obtain information.

SECURITY SPEND MODEL

To quantify budget and increase security spend, you need to demonstrate a return on investment. This means your clients must go beyond people, process, and technology (PPT), and add the value of their information and the location of that information to the equation. Understanding the value of information and the physical location of that information is extremely important in order to align the specific technologies you need to make sure your information is protected. If your intellectual property is worth more than $1 million, it's recommended that you implement tier 2 security technologies to protect it. With today's rapidly growing network infrastructures, ranging from wired to wireless to mobile/4G, the perimeter is yet again being defined. Meaning, with the introduction of mobility it's now bringing threats from the desktop to your pocket with smart devices. This means the traditional perimeter/DMZ is now on person instead of a protected infrasture. However, your investment in tier 2 security technologies will hinge on how you answer the value and information location. This will determine the IT security spend and placement of protection technologies.

SUMMARY

In this chapter, we covered key points that highlight China's strength from an economic perspective and, more importantly, China's ability to acquire technology legally and through industrial espionage. Many detailed cases of China's cyber capability and espionage activity are available on the Internet. Since information on advanced persistent threats is now available in the mainstream media, from the public to the private sector, this situation is now gaining increasing attention, and it's likely that we will see more from China in the decade to come.

References

[1] www.internetworldstats.com/stats8.htm.

[2] www.investopedia.com/terms/g/gdp.asp#axzz20jWyIZ6W.

[3] www.bbc.co.uk/news/business-10792465.

[4] http://bbs.chinadaily.com.cn/thread-739678-1-1.html.

[5] www.economicpopulist.org/content/chinas-industrial-espionage-knows-no-bounds.

[6] www.bloomberg.com/news/2011-12-13/china-based-hacking-of-760-companies-reflects-undeclared-global-cyber-war.html.

[7] www.businessweek.com/articles/2012-03-14/inside-the-chinese-boom-in-corporate-espionage.

[8] www.uscc.gov/hearings/2012hearings/written_testimonies/12_3_26/bejtlich.pdf.

Pawns and Mules

INTRODUCTION

As we have discussed elsewhere in this book, organized and state-sponsored cybercrime consists of numerous "moving parts." The drivers and motives we covered in Chapter 3 explained that acts of cybercrime are primarily driven by a desire for money, but the chapter also discussed the fact that state-sponsored entities also steal intellectual property that has monetary value in terms of technology advancement and quicker time to market. The goal of both the cyber actor, pawn and money mule is rooted in economic and financial gain. With that in mind, this chapter provides insight into how organizations move acquisition targets from point A to point B.

PAWNS IN THE GAME

In the game of chess a pawn is the weakest piece on the board, but oddly enough, if you can successfully maneuver your pawn to your opponent's side of the board you can reclaim any piece that you lost, including the queen, one of the most powerful pieces in the game.

In cybercrime a pawn is a person a cybercriminal recruits to help him or her conduct an attack. In this case, the cybercriminal looks for someone who will fall for social engineering tactics that will get him or her to click on a link or execute a command, thus compromising a system the criminal will use to launch the operation. State-sponsored entities handle their pawns with much precision, and they cultivate a relationship with the pawn very early in the game so that they can get the pawn to do pretty much anything they want. The case described in the following section illustrates how pawns are used.

THE HEARTLAND BREACH

Heartland Payment Systems is a business that processes debit and credit card transactions. At a high level you can think of the company as a clearinghouse for a transaction. According to CardPaymentOptions.com, Heartland is rated as the fifth largest payment processing company in the United States and processes approximately $80 billion in credit card transactions per year. [1] Just in case you're wondering, the largest payment processing company is First Data, which, according to the latest reports, processes every three out of four transactions in the United States. [2] Because these companies process so many credit card transactions, they have been and will continue to be a large target of opportunity among cybercriminals.

Because they process billions of dollars' worth of credit card transactions yearly, these companies have to follow what's called the Payment Card Industry Data Security Standards (PCI-DSS). PCI-DSS is a set of 12 requirements that are supposed to be enforced to ensure the protection and integrity of a credit card transaction in motion and at rest. To ensure that these requirements are enforced, organizations such as Heartland are audited by what is known as a qualified security assessor (QSA), who takes a snapshot of a company's infrastructure which processes and stores credit card transactions. Before the breach occurred, a QSA had certified Heartland as being compliant with PCI-DSS. However, this proved to be useless in stopping the breach. This is because PCI-DSS is essentially Point in Time security (PITs); in other words, even if a company has been PCI-DSS-certified, if the company makes any changes to its security infrastructure after receiving certification these changes can often introduce risk.

According to some sources, in Heartland's case the attackers used an application that passively monitored network traffic and recorded credit card information. [3] In the security industry this is known as a sniffer. At the time of this writing, the installation source of the sniffer had not been identified through open sources. This is not surprising, since sniffers are generally somewhat difficult to detect, especially in large networks. Because sniffers do not transmit data, it is difficult for network security tools to pick them up. (While many security tools exist for finding rootkits and other malware present on a host system, depending on the type of rootkit or malware--and particularly if it is unknown, something commonly referred to as a zero-day and the detection capabilities are somewhat useless.)

All told, according to an article published on Computerworld.com, the cyber-criminals who committed the Heartland breach were able to steal data pertaining to 130 million credit cards, and it cost Heartland approximately $140 million to deal with the resultant fallout. [4]

As security professionals, we often focus on the source code of a vulnerability, which is very important to ensure that the specific vulnerability can be tagged and added to host- or network-based security products. But what happens after the cybercriminal obtains the target (intellectual property, personally identifiable information, credit card information, etc.)? In the case of Heartland, the authorities were able to apprehend only one of the three people responsible for the breach. On March 25, 2010, this person, Albert Gonzalez, was sentenced to 20 years in a U.S. federal prison for his participation not only in the Heartland breach, but also in breaches of many additional companies, among them TJX, according to the U.S. Department of Justice. [5] The charges included wire fraud, war driving, and installation of sniffers on networks, to name a few.

In all, Gonzalez made approximately $2.98 million during his criminal hacking career, according to an article published in *The Independent*. [6] In addition, he was considered to be the leader in the Heartland breach and was reported to have assisted two Russian hackers that were part of a much larger organization. At the time of this writing, these "two Russian hackers" have not been publicly named, nor has the criminal network with which they are associated. The fact that they are reportedly from Russia does not surprise security experts, because when it comes to cybercrime Russia is typically the first place that security experts look. In general, cybercrime rings in Russia are extremely organized. At a high level, they run like a product management organization that brings products or services to market. However, the products and services that these criminal elements bring to bear in the cyber realm are geared toward custom malware and custom social engineering tools that they ultimately are able to successfully release to the public at large in an effort to reap financial gain.

Pawns in the Heartland Case

We began this section by explaining what a pawn is in terms of cybercrime. In the Heartland case, Albert Gonzalez can be considered to be the pawn. According to reports about the case, Gonzalez reportedly suffered from a form of autism as well as Internet addiction, which may have made him an unfortunate target for psychological manipulation and use as a pawn. Although the press made it sound as though Gonzalez was the leader, you can't discount the possibility that he was just a puppet that his Russian handlers were using so as to gain a more powerful foothold in their operation.

Of course, a pawn doesn't have to be a human target/asset; it can also be a computer asset or a business. Although it's unclear in the Heartland case how the criminals actually transported the data, what is known is that to verify that the credit cards were still active they used another pawn: the Odyssey Bar, in eastern Idaho, where they used the cards to make various transactions. [7] The owner of the bar had no idea his bar was being used in this way, and was just as surprised as the legitimate card holders when he began to receive phone calls questioning charges from his bar appearing on the credit bard bills of people who didn't even live in the state of Idaho, let alone the town where the bar was located. The cybercriminals could have used any of a variety of different companies or businesses to verify the validity of the credit cards they stole; however, they chose a small business because small businesses and companies are not typically equipped with sophisticated cyber security countermeasures or fraud detection capabilities.

The Heartland case is a good example of how different types of pawns can be used successfully to commit cybercrime. It's important to emphasize that the goal of a cybercriminal is to not be detected, and we are not suggesting that Heartland was not following security best practices or the QSA was not doing his or her job. The key point that you should take away from this is that this could happen to anyone, and there is no security silver bullet, although security vendors are finding new detection capabilities to close the gaps.

ACQUIRING AND TRANSPORTING STOLEN ASSETS WITHOUT BEING DETECTED

Now that you understand the role that pawns play in helping cybercriminals conduct their attacks, let's take a look at how cybercriminals acquire and transport the assets they've stolen, and how they avoid detection.

Acquiring Assets

Criminals can acquire data in many ways. Here are some examples:

- **SQL injection** This method is very common, and as we discussed earlier in the book, it made headlines for a well-known hacktivist group named LulzSec when they breached the Sony PlayStation network with one SQL injection. A SQL injection can be performed right from a Web browser, and if the server isn't locked down correctly and is susceptible to a SQL injection, it will provide the hacker with not only the type of database software and version being used on the back end, but also table information that might contain sensitive client information.
- **Spear phishing** This is another common tactic that is delivered via e-mail and through social networking (Facebook or Twitter). In this case, the criminal sends the victim a message that's enticing enough to get him or her to click on a link or open an attached file.

- **Social engineering** In this case, the criminal is attempting to gain access or information by pretending he or she is someone else. For example, the criminal could call a corporation pretending he or she is a contracting firm that does business with the corporation, and claim that the firm's computers cannot connect to the corporation's VPN; the criminal can then ask the unsuspecting victim for details on how to connect to the network.
- **Insider threat** The greatest security threat to any organization is someone with access to the front door. The insider threat is real and it's very difficult to even know if someone is leaving with confidential information in his or her briefcase or backpack unless you're checking every bag that leaves the office, and this is not a common practice in the corporate environment.
- **High-flying war driving** This is not as common as driving around in your car looking for an open access point; rather, it involves outfitting a remote control airplane with the right equipment to tap into an open access point and relay that information to a handler who is in close proximity to the signal. Although you don't typically see stories about this in the newspaper, it does happen, and it is a lot more common in organizations that have strong physical security and information of value.

As we mentioned earlier, cybercrime organizations are highly optimized and function like product management organizations. In addition, they have the ability to manufacture custom malware that is specific to the target they want to acquire. Many corporations that have custom-built Internet Web applications are facing even more risk, as they are more likely to have the same if not more vulnerabilities than off-the-shelf software applications. What's even more amazing is that these criminal organizations are recruiting for malware engineers through banner ads offering $2,000 to $5,000 per month. [8] (You shouldn't let this tempt you to give up your day job; if you read the preceding section, you know the sorts of consequences you face if you get caught!)

Transporting Assets

Just as there are several ways to acquire assets, so too are there several ways to transport those assets from the targeted facility. As mentioned earlier, the front door is a company's biggest source of risk. However, the so-called Internet front door of a corporate infrastructure is extremely risky as well, as it will trust just about any connection being established from inside the organization and going outbound to the Internet.

Typically, the following ports/protocols, at a minimum, are wide open from inside the network: port 80 (HTTP), port 443 (HTTPS), port 53 (DNS), and port 25 (SMTP). Additionally, most organizations do not have the tools to categorize and/or identify what applications and file types are exiting their corporate infrastructure. To give you a real-world example, not long ago we were tasked with deploying a data leakage prevention (DLP) device at the perimeter of a major health insurance company. After all the filters were deployed and enabled, the device collected approximately 65,000 events in one hour. Less than 30 minutes after seeing what was exiting and entering the corporate network, however, we were notified that we could no longer view the event data that the device was collecting. The client's corporate policy was very explicit regarding what type of data was allowed to leave the network, and someone was not following that policy based on the information that we viewed. This example is important, as it would probably surprise you how many organizations don't have a clue what type of data is entering or exiting their network. The network and host-based security products that can provide you with such insight/visibility beyond traditional security products can cost in the millions of dollars depending on the size and scope of your network.

Avoiding Detection

Today's sophisticated cybercriminal will avoid being detected at all costs, and will utilize multiple evasion techniques to ensure the target data is successfully moved outside the corporate environment. Here are some examples of typical evasion techniques:

- **Password-protected compressed/encrypted files** One way to evade a data leakage solution is to password-protect a compressed file, as most DLP vendors will not be able to scan the contents of a file that is PGP-encrypted.
- **Commonly known open ports and protocols** As we mentioned earlier, there are typically four ports that have to be allowed in order for any business to conduct Internet operations and receive e-mail. These are typically the only digital transportation vehicles out of a network other than printing it out and walking out the front door with it, or placing the information on a USB device.
- **Applications not supported by corporate policy** Since most corporate environments are not monitoring for the use of TOR, anonymous proxies, Skype, or other Web-based applications.

As you can see, it is very easy for cybercriminals to utilize common applications, ports, and protocols that are legitimate conduits out of a network, but in reality are being used to exflitrate data. However, with the proper detection capabilities put in place you can actually provide more visibility and awareness to what is exiting your infrastructure.

FROM MONEY MULES TO MONEY LAUNDERING

The use of money mules (sometimes referred to as "fraudsters") and money laundering is very common in organized crime, and they work hand and hand. Money mules are typically people who are recruited to perform certain duties, or they can actually be a part of the criminal entity. Their job is to move the product (merchandise or money) themselves, via a third party, or through other organizations located around the world. [9] Money laundering is the act of concealing where the money was obtained from, as the outcome of money laundering makes the money appear to have come from legitimate means. [10]

The Placement Technique

According to the Australian government's Australian Transaction Reports and Analysis Centre, there are three phases to the money laundering process. Placement is the first phase of the process, and it's when the money is first brought into the financial system. [11] Here are some examples of placement techniques:

- **"Smurfing"** Cash from illegal sources is divided among 'deposit specialists' or 'smurfs' who make multiple deposits into multiple accounts (often using various aliases) at any number of financial institutions.
- **Structuring** This involves splitting transactions into separate amounts, each smaller than $10,000.
- **Alternative remittance** This refers to funds transfer services usually provided within ethnic community groups and known by names particular to each culture. Generally such services accept cash, checks, or monetary instruments in one location and pay an equivalent amount to a beneficiary in another location.
- **Electronic transfer** This technique involves the transfer of money through electronic payment systems that do not require sending funds through formal bank accounts. This method is also known as wire transfer.
- **Asset conversion** This involves the purchase of goods. Illegal money is converted into other assets, such as real estate, diamonds, gold, and vehicles, which can then be sold.
- **Bulk movement** This involves the physical transportation and smuggling of cash and monetary instruments, such as money orders and checks.
- **Gambling** This is used to launder money by inserting illegal money into gaming machines. The money inserted can be cashed out and treated as proceeds from gambling. Funds that appear to be winnings can easily be used to justify unusual spikes in income. This income can then be deposited into a legitimate bank account.
- **Insurance purchase** In this case, illegal money is used to buy insurance policies and instruments, which can be cashed in at a later date. The end

result is that the illegal funds have been legitimized by being 'washed' through a legitimate insurance business."

The Layering Technique

The second phase of the money laundering process is known as layering. In this phase, the money is moved, dispersed, and disguised to conceal its origin. [12] Here are some examples:

- **"Electronic funds transfers** Typically, layers are created by moving money through electronic funds transfers into and out of domestic and offshore bank accounts of fictitious individuals and shell companies.

- **Offshore banks** These are banks that allow for the establishment of accounts from nonresident individuals and corporations. A number of countries have well-developed offshore banking sectors. In some cases, these banking sectors follow loose anti-money laundering regulations.

- **Shell corporation** This is a company that is formally established under applicable corporate laws but does not actually conduct a business. Instead, it is used to engage in fictitious transactions or hold accounts and assets to disguise the actual ownership of these accounts and assets.

- **Trusts** These are legal arrangements for holding funds or assets for a specified purpose. These funds or assets are managed by a trustee for the benefit of a specified beneficiary or beneficiaries. Trusts can act as layering tools because they enable the creation of false paper trails and transactions.

- **Walking account** This is an account for which the account holder has provided standing instructions that all funds be transferred immediately on receipt to one or more other accounts. By setting up a series of walking accounts, criminals can automatically create several layers as soon as any funds transfer occurs. Money launderers use this layering technique because it is extremely difficult to detect and money moves very fast through accounts across the world. Due to these reasons, walking accounts create substantial investigation hurdles for regulators.

- **Intermediaries** Lawyers, accountants, and other professionals may be used as intermediaries between the illegal funds and the criminal. Professionals engage in transactions on behalf of a criminal client who remains anonymous. These transactions may include the use of shell corporations, fictitious records, and complex paper trails. Money launderers like to use intermediaries because they lend credibility and decrease suspicion. In addition, these professionals generally have confidentiality obligations to their clients, so the risk of money launderers getting caught is low."

The Integration Technique

In integration, the third phase of the money laundering process, the money is successfully cleaned and looks legitimate; at this point, it is available for use. Here are some common integration techniques: [13]

- **"Credit and debit cards** Credit and debit cards are efficient ways for money launderers to integrate illegal money into the financial system. By maintaining an account in an offshore jurisdiction through which payments are made, the criminals limit the financial trail that leads to their country of residence.
- **Consultants** The use of consultants in money laundering schemes is quite common. The consultant might not even exist. For example, the criminal could actually be the consultant. In this case, the criminal is channeling money back to him/herself. This money is declared as income from services performed and can be used as legitimate funds. In many cases, the criminal will employ an actual consultant (e.g., accountant, lawyer, or investment manager) to do some legitimate work. This could involve purchasing assets. Often, the criminal transfers funds to the consultant's client account from where the consultant makes payments on behalf of the criminal.
- **Corporate financing** This is typically combined with a number of other techniques, including the use of offshore banks, consultants, complex financial arrangements, electronic funds transfers, shell corporations, and actual businesses. This allows money launderers to integrate very large amounts of money into the legitimate financial system.
- **Asset sales and purchases** This technique can be used directly by the criminal or in combination with shell corporations, corporate financing, and other sophisticated methods. The end result is that the criminal can treat the earnings from the transaction as legitimate profits from the sale of the assets.
- **Business recycling** Legitimate businesses that also serve as conduits for money laundering are referred to as 'front businesses.' Cash-intensive retail businesses are some of the most traditional methods of laundering money. This technique combines the different stages of the money laundering process.
- **Import/export transactions** To bring 'legal' money into the criminal's country of residence, the domestic trading company will export goods to the foreign trading company on an overinvoiced basis. The illegal funds are remitted and reported as export earnings. The transaction can work in the reverse direction as well."

Now that you understand the phases of the money laundering process, let's take a look at some real-world examples.

eBay, Craigslist, Autotrader, and Money Laundering

The placement technique of bulk movement has been used to convert stolen money into gift cards that are sold on eBay and Craigslist. This is not to say that all gift cards on eBay are involved in money laundering, but the sheer number of individuals who are selling gift cards on eBay, which is hovering at around 50,000 on average per day, makes it easy for the cybercriminal to blend in.

Meanwhile, the placement technique of asset conversion was in play in the case of Adrian Ghighina, a Romanian national residing in the United States, who acted as a money mule for a Romanian criminal organization and was charged with money laundering through fraudulent online car auctions. [14] Basically, he would offer a car for auction that did not really exist. The operation that he was running netted the criminals approximately $3 million, of which he reportedly received about 40 percent. He also received a 20-year prison sentence.

The ZeuS Botnet and Money Laundering

As we discussed elsewhere in this book, the ZeuS botnet was a malicious program that was targeted at financial institutions' online banking accounts. According to the FBI, the entire ZeuS operation used more than 3,000 individuals acting as money mules, and during the course of the ZeuS operation, the cybercriminals that resided in the Ukraine were able to move approximately $70 million out of the United States. [15] It was also reported that one of the individuals involved with the ZeuS botnet was trying to move money by making deposits at the same bank under multiple aliases, with each deposit not exceeding $10,000 (the placement technique known as structuring). In the United States these suspicious activities fall under the Bank Secrecy Act (BSA) which is an anti-money laundering tool used by all banks within the United States. The following is an example of what U.S. federal banks monitor and report:

- [16]"Currency activity including multiple transactions greater than $10,000.
- Currency activity (single and multiple transactions) below the $10,000 reporting requirement (e.g., between $7,000 and $10,000).
- Currency transactions involving multiple lower dollar transactions (e.g., $3,000) that over a period of time (e.g., 15 days) aggregate to a substantial sum of money (e.g., $30,000).
- Currency transactions aggregated by customer name, tax identification number, or customer information file number."

SUMMARY

The banking system in the United States in highly regulated and highly monitored for acts of nefarious activity such as money laundering. That's why it's surprising to see that most cybercriminals are from Eastern European countries and use pawns as money mules, as they realize the risk they are taking due to the systems that are in place to detect fraud within the borders of the United States. In the case of the Heartland breach, Albert Gonzalez offshored his money to less regulated online financial institutions such as e-gold, which has since been shut down due to many money laundering activities running through its systems. However, given an identity and access to the Internet, there are plenty of unregulated banks through which a cybercriminal can conduct this type of illicit behavior.

In this chapter, you learned that cybercriminals use pawns to move stolen information and to launder money through various online organizations, and that they are not likely to reside in the United States. The Heartland case was important, as it touched on all points that we talked about in this chapter. Additionally, you learned that cybercriminals are extremely sophisticated in terms of creating custom malware for a specific purpose, as they did with the ZeuS botnet that targeted online bank accounts. You also learned that cybercrimes deal with large amounts of money. The few cases listed here netted the cybercriminals close to $300 million. This is big business, but honestly it is not different from how common criminals laundered money before the Internet age. The big difference today is that just about everything is online, and with the right tools you can anonomize your Internet connection and your true identity. Lastly, you learned that the Internet has no geographic boundaries, and this makes it even more difficult to track and recover stolen assets depending on where the cybercriminals reside.

References

[1] http://www.cardpaymentoptions.com/credit-card-processors/heartland-payment-systems-complaints-review-and-rating/.

[2] http://www.cardpaymentoptions.com/credit-card-processors/first-data-review-complaints-and-rating/.

[3] http://www.pciknowledgebase.com/index.php?option=com_content&view=article&id=70:pci-and-the-hartland-payments-breach&catid=19:in-the-media&Itemid=100.

[4] http://www.computerworld.com/s/article/9176507/Heartland_breach_expenses_pegged_at_140M_so_far.

[5] http://www.securityprivacyandthelaw.com/uploads/file/DOJ%20Press%20Release%20Gonzalez%20Sentencing.pdf_.

[6] http://www.independent.co.uk/news/world/americas/hackers-high-life-brought-to-end-with-20year-sentence-1929312.html.

[7] http://www.ktvb.com/news/local/64229557.html.

[8] http://krebsonsecurity.com/2011/06/criminal-classifieds-malware-writers-wanted/.

[9] http://www.lloydstsb.com/security/money_mules.asp.

[10] http://www.nvo.com/beaulier/whitecollarcrime/.

[11] http://www.austrac.gov.au/elearning/pdf/intro_amlctf_placement_techniques.pdf.

[12] http://www.austrac.gov.au/elearning/pdf/intro_amlctf_layering_techniques.pdf.

[13] http://www.austrac.gov.au/elearning/pdf/intro_amlctf_integration_techniques.pdf.

[14] http://news.softpedia.com/news/Romanian-Man-Admits-Involvement-in-Fake-Online-Auc-tions-Scheme-185109.shtml.

[15] http://www.pcworld.com/article/206820/five_arrested_in_scam_involving_theft_via_bot-net.html.

[16] http://www.ffiec.gov/bsa_aml_infobase/pages_manual/OLM_015.htm.

CHAPTER 9

Globalization: Emerging Markets Aren't Just for Traditional Investors Anymore

INFORMATION IN THIS CHAPTER:

- Introduction

CONTENTS

INTRODUCTION

The concept of emerging markets comes from the field of economics during the 1980s, in a search to define a group of nations that are in the process of rapid growth and accelerated industrialization. Whether they are defined as BRIC (Brazil, Russia, India, China)[1], or by the larger list espoused by Dow Jones,[2] the concept of an emerging market is one of high business risk economically, where risk can be positive or negative. While multinational companies willing to invest in these markets can potentially expect higher than average growth, there are possible random uncertainties, including political instability, which can shut down the country's ability to deliver results, often with little to no notice. These emerging markets are also a potential boon for cybercrime today. The rapidly expanding attack surface in these countries, and the budding disposable income readily available for purloining, provides a perfect storm for these new-age criminals.

Uncertainty, Instability, and the Potential for Illicit Gain

This uncertainty has, at various times, been exhibited in nation states such as the former Union of Soviet Socialist Republics (USSR), now known as the Russian Federation.[3] These key emerging markets can therefore deliver excellent results to those who can stomach the potential emotional rollercoaster, but they can likewise cause a shutdown of activities for three to six months

Dow Jones - Emerging Markets[2]

- Argentina,
- Bahrain,
- Brazil,
- Bulgaria,
- Chile,
- China,
- Colombia,
- Czech Republic,
- Egypt,
- Estonia,
- Hungary,
- India,
- Indonesia,
- Jordan,
- Kuwait,
- Latvia,
- Lithuania,
- Malaysia,
- Mauritius,
- Mexico,
- Morocco,
- Oman,
- Pakistan,
- Peru,
- Philippines,
- Poland,
- Qatar,
- Romania,
- Russia,
- Slovakia,
- South Africa,
- Sri Lanka,
- Thailand,
- Turkey,
- United Arab Emirates

during political or economic unrest. In instances such as the ongoing economic unrest in Greece, for example, the possibility in June 2012 of that country's exit from the euro zone (at a price tag estimated at more than $1 trillion!) had policymakers the world over consumed with worry that in letting go of Greece, the risk posed to social and political stability in Europe would be immense. The ramifications, if this were to occur, would be mind blowing, ranging from hyperinflation, to political instability, to the erosion of public services, to an increase in legal and illegal migration across borders. All of these fears were valid given the tenuous state of the country in question. When viewed in an economic context, along with the potential threat poised in the form of social breakdown, the result may very well have been the collapse of the rule of law and order within Greece itself. This collapse, it was feared, would allow for a vacuum to be filled by nationalists seeking establishing themselves in positions of power in order to further their agendas. What would Greece's neighbors do or think should the instability described above come to pass? During this time, Anonymous, the online hacktivist group, was engaged in operations against the Greek government presumably on behalf of the people of Greece. Instability in a nation or region where, due to economics, stability is strained at best can quickly teeter, thus the challenge in pursuing business in emerging markets.

The Jasmine Revolution

The civil unrest seen in North Africa during the now infamous 'Jasmine Revolution'[4] is an excellent example of how uncertainty in emerging markets can influence (negatively and positively) the potential for business to be conducted. This revolution—perhaps one of the most pivotal of such actions in modern times—had its roots in Tunisia. Figure 9.1 shows a map of Tunisia, highlighting key areas of activity and populace. Prior to the inception

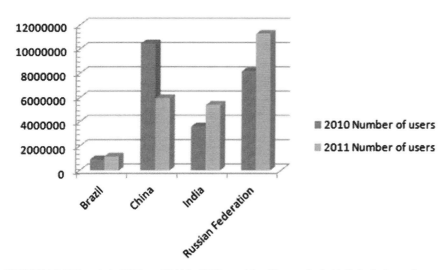

FIGURE 9.1 Threats in 2010 vs. 2011 in BRIC countries (Kaspersky Lab). Note that a major legal reform in regard to cybercrime and intellectual property in China caused a drastic decrease (and potential off-shoring) of malware in early 2011

of the Jasmine Revolution,[5] Tunisia was a nation of approximately 10.4 million people, the majority of which spoke both Arabic and French. Islam was the state religion, though the government supported a secular society—an idea not universally held in the region. The majority of Tunisia's economy was derived from a rather profitable tourist industry. Its agricultural contributions were composed of mostly olive oil (a staple in the Mediterranean region) and oranges. It had a moderate unemployment rate (~14 percent), though it saw higher than average rates of joblessness among its educated youths and within the internal, rural areas away from the coast of the Mediterranean.

On December 17, 2010, something of monumental note impacted Tunisia, the region, and the world over. A 26-year-old computer science graduate and fruit vendor named Mohamed Bouazizi set himself ablaze after experiencing what has been referred to as police brutality on the streets of Sidi Bouzid. The altercation was initiated when a female police officer slapped Bouazizi and ordered him to pack up his vendor cart due to his inability to procure a license to vend on the streets.[6] For Bouazizi, this was the final straw in a growing list of grievances that included not being able to find employment in his chosen field and having to resort to selling fruit to support his seven siblings. Some 18 days later, the young man died in the hospital due to his injuries. Bouazizi became an unwitting martyr for what would become a pivotal social revolution in Tunisia. The movement quickly took off in other North African nations, such as Tunisia's neighbor to the east, Libya, where the regime of Muammar al-Gaddafi was toppled, followed by perhaps the most volatile manifestation of the revolution in Egypt, which saw the end of the Hosni Mubarak regime.

The world waited and watched with bated breath as regimes and despotic dictatorships fell one after another. There was great concern as to the potential implications that these changes in stability would have on the production of regional goods and, of course, the global production and availability of crude oil. One year after the Jasmine Revolution, Tunisia was, in some respects, worse off economically than it was before. Tourism, the nation's largest source of income in the form of foreign currency, had fallen by more than 50 percent. Foreign direct investment in the nation had fallen by more than 20 percent, and more than 80 percent of foreign corporations had elected to leave the country. Labor markets within Tunisia continued to worsen, with a heightened number of layoffs and the return of migrant Tunisian workers from Libya due to that country's own ensuing revolution. The number of unemployed one year after the Jasmine Revolution had risen from 500,000 to 700,000 (an increase from 14 percent to 17 percent).[7]

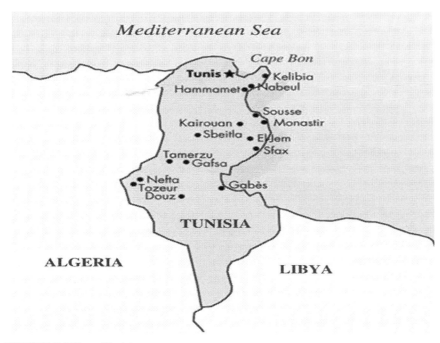

FIGURE 9.2 Map of Tunisia

The potential for instability in emerging markets poses challenges for legitimate businesses. Exxon-Mobil, for example, noted in February 2011 that the ramifications of the Jasmine Revolution in North Africa and beyond would impact its ability to continue to grow at rates seen in previous years.[8] It could be argued that the presence of instability in geographies noted as being emerging markets can and often does lead to the growth and development of illicit markets and privateers seeking to profit by any means they can.(see Figure 9.2)

Latin America: The Cybercriminal's Paradise

Along the same lines, countries such as Panama and Costa Rica[9] were often considered "fiscal paradises" due to the lack of laws and regulations prohibiting illegal or illicit activity. Many emerging markets have now become "Black-Hat" paradises in the eyes of those who are intent on doing harm in the cyber world. The lack of international frameworks for enforcing cybercrime has often created these cybercrime havens in some emerging market countries. It would be fair to compare these former "off-shore financial centers," which manifested themselves as "economies with financial sectors disproportionate to their resident population (International Monetary Fund) (International Monetary Fund)",[10] to the new cybercrime-friendly countries, which manifest themselves with digital penetration and sourced attack vectors disproportionate to their resident population.

Costa Rica[9] is an example of what a lack of laws or legal frameworks that prohibit illicit or illegal cyber activity can result in. Costa Rica was a favored nation for organized crime enterprises in the United States.[11] Several factors influenced this, not the least of which was the pressure that organized crime syndicates felt over time from the U.S. Department of Justice (DoJ) with respect to illegal gambling operations. Perhaps the best-noted and chronicled example of this is the case that Joseph Menn outlined in his book, *Fatal System Error: the Hunt for the New Crime Lords Who Are Bringing down the Internet*. In his book, Menn follows the story of Barrett Lyon, a young entrepreneur who was contracted by an offshore Internet-based gambling operation to provide consultancy services. The relationship that ensued saw Lyon eventually launch a business with angel funding from his clients (who he later found out were members of organized crime families in the United States) while also tracking down Russian cybercriminals responsible for advanced denial of service (DoS) attacks against his clients and other operations.[12] It should be noted that in other parts of Latin America—Brazil,[13] for example—the hacker mindset is alive and well in addition to being homegrown. Brazil has, perhaps more so than any other Latin American nation, save Mexico, the largest (per capita) population of native-born hackers. Many of these hackers are dedicated "WhiteHat" hackers working diligently to secure what is broken and prevent further exploitation by those with malicious intent. However, there are equally as many who strive to profit by any means necessary, targeting vulnerable environments within Brazil and beyond its geographic borders.[14] In direct contrast, the new wave of malware being created in Argentina and Peru, both of which are defined as tier 2 emerging countries, has far exceeded the malicious nature previously found in American, Brazilian, and even Mexican malware. Argentina's malware has followed the BYOD wave successfully, and that country's cybercriminals have created some very successful mobile malware.

Malware Presence in Emerging Markets

Following the list of emerging markets as defined by Dow Jones, the authors have reviewed the number of attacks, using data provided by Kaspersky Lab for 2010 and 2011. The BRIC countries show a definite increase in malware attacks. (see Figure 9.3)

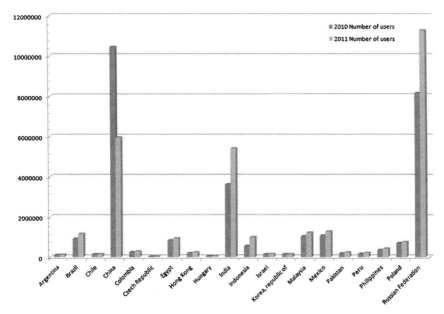

FIGURE 9.3 Emerging Markets Malware 2010 vs. 2011 (Kaspersky Lab)

The Evolution of Law and Policy in Fighting Cybercrime in Emerging Markets

As the monetary policies have evolved, so will eventually the cybercrime legislation in these countries though today.[15] Eventually, there will be greater success with groups such as IMPACT[1] that will help create an international cybercrime law equal to maritime law, which will provide equal footing in the enforcement of cybercrime. Interpol recently announced a "law enforcement tech geek heaven" in Singapore. The INTERPOL Global Complex for Innovation will function as an R&D lab, training facility, and forensics lab for all things cybercrime, and is slated to open in 2014.[16]

As Michael Moran, INTERPOL's Acting Assistant Director for Cyber Security and Crime, succinctly explained, "Most cybercrime-investigating cops worldwide had inadequate budgets, overwhelming workloads, and talent problems".

---

[1] IMPACT (International Multilateral Partnership Against Cyber-Terrorism) is the world's first multilateral, public-private sector collaborative institution against cyber-terrorism. IMPACT serves as a global platform bringing together governments of the world, the industry and academia to enhance the capability of the global community to prevent, defend and respond to cyber threats.

As Moran put it, "recruiting long-haired geeks is not easy for law enforcement." The Interpol website well states that

> "The use of the Internet by terrorists, particularly for recruitment and the incitement of radicalization, poses a serious threat to national and international security."

The Changing Nature of Cybercrime

In the past, individuals or small groups of individuals have committed cybercrime. This worked well for many and was a scalable model during the infancy of the Internet. During the first decade of the 21st century, we saw miscreant activity subside and give way to loose confederations of individuals seeking to profit from the vulnerability of others. Nowhere was this truer than in the shadowy underground that surrounded credit card fraudsters. Sites and forums dedicated to the trafficking, exploitation, and fraudulent use of credit card sites, such as those run by the infamous Shadowcrew,[17] [18] others such as DarkMarket, and perhaps most notably, CardersMarket,[19] run by Max Ray Vision (a.k.a. Max Ray Butler), for example, gave birth to a new era of cybercriminal activity. These sites acted as information forums, clearinghouses, and tutorial environments for mature and aspiring fraudsters bent on exploiting as many unsuspecting individuals as possible.

Today we see an emerging trend with traditional organized crime syndicates and criminally minded technology professionals working together and pooling their resources and expertise. This approach has been very effective for the criminals involved. In 2007 and 2008 the cost of cybercrime worldwide was estimated at approximately USD 8 billion. As for corporate cyber espionage, cyber criminals have stolen intellectual property from businesses worldwide worth up to USD 1 trillion."[20] Meanwhile, countries like Pakistan, Afghanistan, and Kyrgyzstan, which have relatively small populations, will continue to have high incidence of source vectors for attacks. Similarly, although China's population is extremely large, the internet penetration is relatively small and extremely controlled.

Interestingly enough, the attacks we see from China are usually highly structured, and exhibit military tactics in their development, but more on this subject can be found in Chapter 7. As early as 2010, Symantec pointed out in their Global Internet Security Threat Report that criminal activity is migrating from mature markets to emerging economies. Perhaps, it intimated, awareness and enforcement are even more lax in those countries.[21] Symantec called out specific emerging countries like Brazil, India, Poland,

Vietnam and Russia. During 2009, these emerging economies moved up the rankings as a source and target of malicious activity by cybercriminals. Symantec, in their findings, suggest that government crackdowns in developed countries may have led cybercriminals to launch their attacks from these countries from the developing world, where they are less likely to be prosecuted.

In the 2010 report, Symantec highlighted that victims of cybercrime that occurred in emerging markets (Brazil, India, China, Poland, Mexico, UAE), were more prevalent, 80%, versus those that occurred in developed markets (UK, US, France, Germany, Italy, Australia, Canada, Sweden, Japan, Spain, Netherlands, Denmark, Belgium) 60% The same report highlighted an interesting geographical distribution, where the computer virus and malware crime capitals within the group of cities included in the study are: Mexico (71 percent), Brazil and China (68 percent) and South Africa (65 percent), all major economies within emerging countries.[21]

Cyber criminals have taken and continue to take advantage of Emerging Market countries in a variety of ways. They are often used as transition points (jump-off points) or as pass-through countries for "muling", and at times even staging areas for their crimes. Even though many crimes are intended against targets in mature markets, because they either initiate, or hop through an emerging market, means that the jurisdictional issues are convoluted at best and impossible at worst. The fact that many emerging market countries have no developed cyber-crime legislation means that cyber-criminals can traipse through these countries with little to no risk of being caught. In those countries that have enacted cyber-crime legislation, the often laughable enforcement or high corruption index means that the cyber-criminals can calmly ignore the laws as though they did not exist, or simply buy their piece of mind and way out of jail with some of their ill-gotten moneys.

The very nature of the internet as a global and ubiquitous vehicle for communication and research means that it is similarly well suited as a global platform for cyber-crime. This globalization also means that proximity, whether in time or space, is irrelevant in communication as in cybercrime. As such, the attacker not only need not be close to their victim, but need not be around at the time of the actual execution of the crime. In the same way that an email can lie dormant for days before being read, a spammer can send his millions of infected e-mails, and the recipients can be in hundreds of countries, without having to even carry a passport. As well, his victims can open the emails immediately, or wait months to do so. Regardless of the eventual action by the victim, the perpetrator can even have deceased, and the crime is still carried out, and potentially monetized. Likewise, a botnet can become a very effective tool in an internationally delivered Distributed Denial of Service (DDoS), and the target

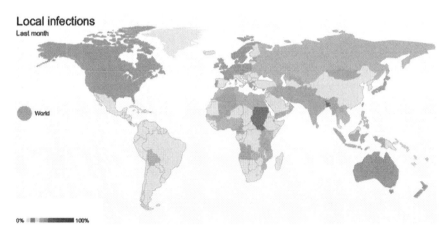

FIGURE 9.4 Kaspersky Lab – SecureList Showing Local Infections during January 2012[22]

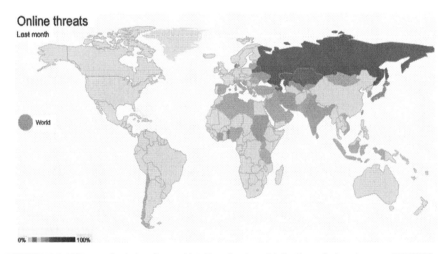

FIGURE 9.5 Kaspersky Lab – SecureList Showing Local Infections during January 2012[22]

need not be anywhere near the attacker, and could even have multiple middle-men in the transaction.

As you can tell from a quick glance at the map in Figure 9.4, local infections are low in mature market countries. By comparison, when we switch our view to online threats in Figure 9.5, we see that mature markets are a target in many online threats. The simple conclusion is that these threats are initiating in emerging market countries, and targeting mature markets.

References

[1] Foroohar, Rana. Power Up. The Daily Beast. March 20, 2009. http://www.thedailybeast.com/newsweek/2009/03/20/power-up.html.

[2] Dow Jones – Total Stock Market Indexes. http://www.djindexes.com/mdsidx/downloads/brochure_info/Dow_Jones_Total_Stock_Market_Indexes_Brochure.pdf.

[3] Bandow, Doug. The Baltic States Prosper Amidst the Euro's Ruin. http://cxssr.org/2012/06/the-baltic-states-prosper-amidst-the-euros-ruin/.

[4] http://prospectjournal.ucsd.edu/blog/index.php/a-simplified-timeline-of-the-jasmine-revolution/.

[5] Brookings. How will Tunisia's Jasmine Revolution Affect the Arab World? http://www.brookings.edu/research/opinions/2011/01/24-tunisia-assaad.

[6] Feist, William and Andrade, Dominic. Jasmine Revolution's economic consequences. New Jersey Newsroom. March 9, 2011.

[7] Achy, Lahcen. Tunisia's Economy One Year after the Jasmine Revolution. http://carnegieendowment.org/2011/12/27/tunisia-s-economy-one-year-after-jasmine-revolution/8z5s.

[8] NASDAQ. Assessing Jasmine Revolution's Impact on Exxon Mobil Production. http://community.nasdaq.com/News/2011-02/assessing-jasmine-revolutions-impact-on-exxonmobil-production.aspx?storyid=59421.

[9] Grant, Jason. The Star-Ledger. Feds in N.J. bust illegal online gambling ring. http://www.nj.com/news/index.ssf/2012/05/feds_bust_illegal_online_gambl.html.

[10] Zoromé, Ahmed. Concept of Offshore Financial Centers: In Search of an Operational Definition. http://www.imf.org/external/pubs/ft/wp/2007/wp0787.pdf.

[11] The Costarican Times. http://www.costaricantimes.com/la-cosa-nostra-implicated-in-operating-illegal-costa-rican-web-gambling-site-beteagle-com/741.

[12] Leyden, John. The Register. http://www.theregister.co.uk/2010/03/01/fatal_system_error_book_review/.

[13] Wilson, Tim. Latin America's Criminal Gangs Get Tech-Savvy. http://www.insightcrime.org/insight-latest-news/item/1976-latin-americas-criminal-gangs-get-tech-savvy.

[14] Securelist. http://www.dataprotectioncenter.com/antivirus/kaspersky/brazilian-trojans-beyond-borders/.

[15] Navarro Isla, Jorge. Cyber regulation in Latin America. http://www.cepal.org/socinfo/noticias/paginas/0/30390/newsletter15ENG.pdf.

[16] http://www.fastcompany.com/1814963/inside-interpols-new-cybercrime-innovation-center.

[17] http://www.businessweek.com/stories/2005-05-29/hacker-hunters.

[18] http://www.theregister.co.uk/2006/06/30/shadowcrew_sentencing/.

[19] http://www.wired.com/techbiz/people/magazine/17-01/ff_max_butler.

[20] http://www.interpol.int/Crime-areas/Cybercrime/Cybercrime.

[21] Symantec – Internet Security Threat Report, Volume 17. http://www.symantec.com/business/theme.jsp?themeid=threatreport.

[22] Kaspersky Lab. Secure List. Local Infections. http://www.securelist.com/en/statistics#/en/map/oas/month.

CHAPTER 10

America, Land of Opportunity

INFORMATION IN THIS CHAPTER:

- The Birth of the Modern Internet
- When Purpose is Corrupted by Conflicting Intent
- Defining the Norm: The Era of the Cybercriminal on America's Internet
- Locking the Doors while Opening the Windows: Inviting the Cybercriminal into Our World and Our Lives
- Does Education Decrease Cybercrime in the United States?
- Industrial Espionage and the American Experience
- When the Crime is Not Motivated by Economics

CONTENTS

INTRODUCTION

Since the dawn of the Advanced Research Projects Agency Network (ARPANET) in Hawaii on November 21, 1969 there were two key purposes for the communications network that evolved into the Internet. The first was to ensure that in the event of a global nuclear conflict, military communications would be maintained, allowing for retaliatory actions to ensue. The second purpose was as an adjunct communications mechanism useful in academic research. As such, the program that gave birth to what is referred to as the "Internet" was initially conceived and funded by a U.S. military research organization, the Advanced Research Projects Agency (ARPA), which later became the Defense Advanced Research Projects Agency (DARPA). It was under this second purpose that the Internet grew exponentially during the 1970s and 1980s. Examples of this growth can be seen in Internet host count histories assembled by organizations such as the Internet Security Consortium.[1] In a study it conducted, host counts from 1981 through 2012 being advertised and thus noted through analysis of the Domain Name Service (DNS) protocol have grown over 4,170, 038 times![2]

In this chapter, we will explore how the reality of the development of the Internet as an American military project, and its commercialization have created the perfect storm for hackers.

THE BIRTH OF THE MODERN INTERNET

Due to the Internet's two earliest primary use cases and their often conflicting needs, the network was eventually segmented into two unique communications environments: MILNET and NSFNET. These two environments would later give way to what is now known as the modern Internet. In order to achieve fast, consistent communications over the Internet, security was often sacrificed. Fast speeds were difficult to achieve due to the geographic and platform disparity that comprised the Internet at the time, and efficient communications were difficult to achieve even under the best conditions where fault-tolerant, redundant infrastructures were established and maintained by experienced engineers in order to address fragmentation and other anomalous network behavior including latency, jitter, and delay. Secure communications for those not involved in military or government contexts was, to a greater or lesser extent, an afterthought and would later prove to be not only required but absolutely essential in trying to stave off threats of all types. And these early environments were often considered and viewed to be equal to "private clouds" due to their lack of participants and notoriety.

WHEN PURPOSE IS CORRUPTED BY CONFLICTING INTENT

Because one of the main purposes of the Internet was academic, most of the nodes that were launched were stood up within university campuses. This created a secluded environment that was intended for sharing and collaboration among academic researchers, such as that seen at the University of Illinois National Center for Supercomputing Applications (NCSA) whose mission was to provide powerful computers and expert support, to help thousands of scientists and engineers across the country improve the world.[3] Information security, though important, was not a primary concern, if a concern at all, as researchers and academics required access and insight to knowledge and information contained within private and semiprivate enclaves attached to the Internet.

Despite the purity of this academic idealism, the data made available was often of considerable worth to not only the researchers and academics pioneering new technological innovations, but also those with less altruistic intentions. In one noteworthy case, a team of industrious East German hackers compromised the

Lawrence Livermore National Laboratory (LLNL) in 1986.[4] In August of that year Clifford Stoll, an employee of LLNL, was asked by his then supervisor Dave Cleveland to resolve what at the time appeared to be an anomalous accounting error stemming from a 75-cent charge to the computer usage accounts. Like most research labs LLNL took accounting matters seriously, and though the error was low in terms of dollar amount, it had to be addressed to ensure that funding continued and scrutiny subsided. Stoll was able to trace the error to what appeared to be an unauthorized user who had apparently used approximately nine seconds of computer time and had not paid for it, thus explaining the 75-cent error. However, Stoll soon realized that this was no ordinary case of unauthorized access, but rather a case where a "cracker," or what we now refer to as a "hacker," had acquired root access to the LLNL system he managed by exploiting a movemail function vulnerability in the original GNU Emacs.

Stoll spent a lot of time and energy tracing the origins of the hackers involved in the compromise, and eventually discovered that the infiltration was taking place via a 1200-baud modem connection. Soon he engaged the help of colleagues Paul Murray and Llody Bellknap to help verify and trace the phone lines in question. Over the course of a long weekend, through deductive reasoning and analysis, Stoll and his colleagues tracked the intrusion to a line being routed to LLNL from the Tymnet routing service which eventually led the team to a call center at MITRE, a defense industrial base contractor located in McLean, Virginia.

Over time, Stoll noted and observed the hacker's activity as he sought to exploit U.S. military installations looking for intelligence linked to nuclear weapons and other technology/intelligence that would be of interest to what appeared to be a foreign intelligence service. In many instances, as Stoll would go on to note in his now classic work *The Cuckoo's Egg: Tracking a Spy Through the Maze of Computer Espionage,*[5] that the hacker would gain entrance without using any credentials at all; they simply logged on as guests of the system. As Stoll continued his investigation he contacted the FBI, CIA, NSA, and U.S. Air Force OSI to ask for their help and guidance while also alerting them to the issue.

Stoll's investigation eventually led him to conclude that the intrusion was originating in West Germany via a satellite connection. After some time and effort, the communications trail led to a university in Bremen. Stoll worked to establish what in many respects was the first known instance of a *honey pot*, which eventually saw the hacker reveal himself over time. Stoll prepared an intricate set of false data that would be of potential interest to the hackers, and used it to lure them into deeper hacking into the server so that he could analyze and comprehend their tactics. Markus Hess, the hacker in question, had for many years been selling illegally gained intelligence to the Soviet KGB. Information leading to Hess's subsequent arrest and trial was acquired via the honey pot project stood up by Stoll in which a Hungarian intelligence operative had tried to establish contact with a fictitious entity tied to the honey pot designed by Stoll. This information could only have been provided by Hess through his illegal access to systems controlled by Stoll at LLNL.

Stoll's work continues to be cited as a best practice, and should be required reading by security experts engaged in counterintelligence, digital forensics, and incident response. Many believe that Stoll's experiences as chronicled in his book were the first true example of criminal hacking noted and publicly disclosed, thus, its importance here and throughout this book as we continue to see and example after example of activity such as this on a global scale.

As Internet connectivity became ubiquitous and bandwidth increased to more and more households and organizations in the United States, the lack of inherent and integrated security in the primary communication protocols provided a perfect playground for hackers as first noted by Stoll. As consumers supplanted academics as the primary customers of the Internet-based communications services the anonymity of the medium created a perfect forum for illicit Web sites containing myriad materials of an illegal and black market nature, including, but not limited to, illegal software keys and licenses (known as "warez" sites), pornographic materials, music, and much, much more. The net effect is an uncontrolled international marketplace.

DEFINING THE NORM: THE ERA OF THE CYBERCRIMINAL ON AMERICA'S INTERNET

As the digital era began and the percentage of connected households and commercial entities increased, this ubiquitous connectivity also created a well-defined demographic for cybercriminals. Early adopters of the new commercial Internet—whether for e-mail, pornography, warez and other illicitly gained materials, and eventually the Amazon digital mall—were people with disposable income, with little concern for security, and with available time to surf at dial-up speeds. Those who have never heard the magic warbling of a modem connecting over a dial-up connection, or never had to remove the wall plate at a hotel to jerry-rig a modem connection, can only compare the experience to trying to navigate the World Wide Web using a 1-inch screen and text-based menus. Despite the instability of the infrastructure, the noise on the line, and the frustration of dropped calls whenever Mom picked up the line, those who were wired and on "the Net" were the coolest cats. They were, literally, plugged in to the source of information and were experiencing the world in a way that had not previously been thought possible.

This commercialization was greatly accelerated by improvements made to the Hypertext Markup Language (HTML). HTML enabled a graphical interface that made the Internet and the World Wide Web friendlier and more accessible. This accessibility and user-friendly experience saw the transition of the Internet from academia to mainstream America. The University of Illinois at Urbana-Champaign led the way with the MOSAIC project,[6] [7] leveraging the work done at CERN[8] in Europe. You no longer needed to be limited to text, and to strange and esoteric commands. Thanks to the efforts of those at CERN and the University of Illinois (and later the Mosaic/Netscape corporation), you could now select an image, drag it over to your desktop, and it would download. When Sandra Bullock's character ordered a pizza online in the movie *The Net* in 1995, it foretold a drastic change to commerce as we knew it. One day, we too would be able to order a pizza without having to endure the condescending tone of the order taker. The magic of Apple iTunes synchronizing purchases across multiple devices within seconds of a micro transaction is the true measure of the advances that have taken place in the past 45 years. The lack of security, however, threatened to slow down the economic aspects of the new Internet.

With this came an extreme form of egocentrism that evolved from a feeling that the United States owned the Internet. Despite it being originally developed as an international academic network, a large part of the footprint of the Internet in the early days was hosted in the United States, where connectivity was better and the infrastructure more robust. Even today, a large percentage of traffic passes through servers hosted on U.S. soil. Some of this egocentrism can be attributed to the fact that the original funding for the Internet came from DARPA, but we don't see such pride over other interesting DARPA inventions such as Velcro.

While other countries were expected to utilize country-specific top-level domains (TLDs), such as .co.uk for the United Kingdom and .com.co for Colombia, the great majority of U.S. Web sites are under the .com TLD, and not .com.us. The egocentrism of Americans as it relates to the Internet is such that studies show the great majority of the U.S. Internet population does not even know there is a .us TLD.

LOCKING THE DOORS WHILE OPENING THE WINDOWS: INVITING THE CYBERCRIMINAL INTO OUR WORLD AND OUR LIVES

The combination of a large Internet population and consumer ignorance creates a prime target for cybercriminals. They now have a soft target that they can easily con and exploit. As the percentage of Internet users grew in the United States, so did cybercrime.

Between 2001 and 2009, for every additional percentage of Internet users, the monetary impact of cybercrime in the United States has grown by

Table 10.1 Trends of US Population, Internet Users, and Financial Impact of Internet Crime

Year	Population	Users	Losses in Millions	% Pop.	Broadband	Usage Source
2000	281,421,906	124,000,000		44.10%	n/a	ITU
2001	285,317,559	142,823,008	17,000,000	50.00%	n/a	ITU
2002	288,368,698	167,196,688	54,000,000	58.00%	n/a	ITU
2003	290,809,777	172,250,000	125,600,000	59.20%	n/a	ITU
2004	293,271,500	201,661,159	68,000,000	68.80%	n/a	Nielsen Online
2005	299,093,237	203,824,428	183,120,000	68.10%	n/a	Nielsen Online
2007	301,967,681	212,080,135	239,090,000	70.20%	n/a	Nielsen Online
2008	303,824,646	220,141,969	264,600,000	72.50%	n/a	Nielsen Online
2009	307,212,123	227,719,000	559,700,000	74.10%	n/a	Nielsen Online
2010	310,232,863	239,893,600		77.30%	85,287,100	ITU

$22.5 million dollars. We can clearly see that consumer ignorance influences the success of the cybercriminal (see Table 10.1).

What we see is that not only does the number of victims increase, so does the average cost of each crime. Many of these crimes have "migrated" online. Fraudsters that previously would act through personal interaction and telephone media now use the Internet. However, this does not necessarily mean all crime has gone digital. There are still household robberies, vandalism, and carjackings. To quote a recent conversation with Eugene Kaspersky, the founder and CEO of Kaspersky Lab, "We certainly do not see criminals taking up C++ classes while in prison to digitize their next crime wave(see Figure 10.1)." (Kaspersky, 2012)[10]

Cybercrime Goes Mainstream

We do, however, start to see an interesting evolution in U.S. cybercrime as the Internet becomes increasingly mainstream. As more people have migrated their communications online, we have seen some criminals migrate their vehicles of crime online as well. The first of these, the chain letter of old, becomes a chain e-mail. Whereas these previously arrived by mail and were usually ignored, they start to arrive by e-mail, where they are equally ignored. In fact, according to Postini, a company owned by Google, today more than 94% of e-mail traffic generated online is considered spam. And spam, while a global problem, is

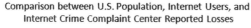

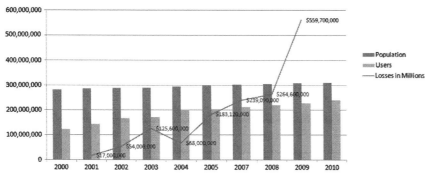

http://www.ic3.gov/media/annualreports.aspx

FIGURE 10.1 Comparison of Growth in U.S. Population, Internet Users, and Internet Crime Economic Impact [9]

particularly bad in U.S. Internet traffic (see Figure 10.2), where it has become a problem worthy of legislation.

Historically, the earliest mail-borne viruses, such as MELISSA and I LOVE YOU, proved to be substantially more damaging, mainly because of the lack of protection users had at the time. With better intelligence and protection, we now know that 80 percent of the spam traffic in the United States and Europe can be attributed to 100 known spam operations. Out of these, 62 of the largest offenders are located within the United States, meaning that the greatest part of this traffic is created, transported, filtered, and rejected all within U.S. servers. External countries responsible for spam traffic are the Russian Federation, with 11 known offenders, and Canada and India, each with five[12] (SPAMhous. org). These statistics are difficult to pinpoint to specific countries, since many of these spam offenders, in order to maintain their business, often change networks, providers, and countries. Still, with such a great percentage of the spam offenders within the United States, e-mail traffic statistics and the percentage of blocked traffic continues to threaten growth of the commercial infrastructure of e-mail. Users increasingly become wary of e-mails, and refuse to provide their e-mail addresses, concerned that they will be infected by those undesirable messages. Early on, it was not like this. E-mail was such a rarity that each message

> Eighty percent of spam received by Internet users in North America and Europe can be traced via aliases, addresses, redirects, locations of servers, domains, and DNS setups, to a hardcore group of around 100 known spam operations, almost all of whom are listed in the ROKSO database. (SPAMhous.org) (http://www.spamhaus.org/rokso/)

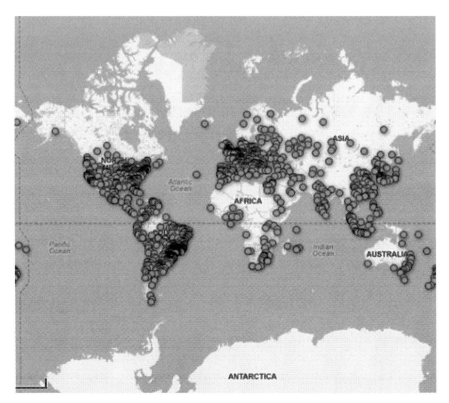

FIGURE 10.2 Where Spam Was Being Created in July 2012, According to Postini [11]

was read and reread with the anticipation of a love letter, even if the content was usually substantially less intriguing. Users of the earliest mail systems, such as CompuServ, would log on just to see if an expected message had arrived. With the amount of spam on the rise, we now approach our mailbox with the same fear as a bomb technician approaching an abandoned suitcase at an airport.

Spamming and Phishing

With the growth of e-mail volume, it became important for the criminal aspect to be able to break through the barriers that users and corporations started to erect. When e-mail became mainstream, the mere number of users became a desirable target for scammers. We saw ISPs and companies rush to protect users from gullibly falling prey to them. In 1997, the first Real-Time Black List was published, looking to solve the root cause of the problem.[13] (SecureList). As users became more discerning about the spam issue, spammers also developed new techniques to overcome these filters. At first, simple content filters could find strings of words to be filtered, such as curse words, the word *Viagra*, and other terms. As these started to be filtered, spammers started to look for inventive ways to spell the same words, such as *VLAGRA*, then *V14Gr4*, and eventually migrated to images of the word so as to bypass these rudimentary filters.

E-mail traffic is responsible for a large percentage of the traffic on the Internet backbone. As seen on Figure 10.3, this growth has been explosive in the last 5 years. Unfortunately, as traffic increases, so does malware.

In the last year alone, we see that the threats found by Kaspersky Lab, as shown in Figure 10.4, have reached more than 600,000 attacks per year.

Phishing messages, intended to guide a user to a particular site where the user's information could be gathered, also became a favorite tool of the more discriminating spammers, who had begun to develop new and innovative ways to reach users, and to trick them into falling for their schemes. While phishing became known as a mass problem, the criminal element evolved into spear phishing and whale phishing, two interesting variations on the theme. In spear phishing, rather than casting a wide net, spammers strategically focus on specific users they want, and send directed attacks. In whale phishing, entire operations are focused on specific high-level executives of a corporation.

It is with directed and strategically crafted attacks such as these that RSA became victim of an attack that threatened its SecurID technology. A carefully crafted message, with an Excel attachment with the name "2011 Recruitment plan.xls," was sent to specific individuals within the company. A single employee that

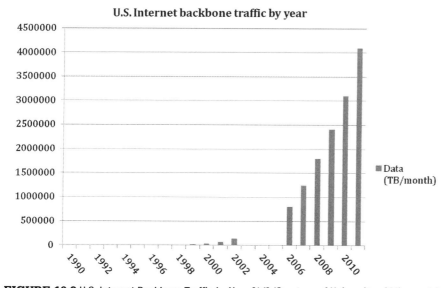

FIGURE 10.3 U.S. Internet Backbone Traffic by Year [14] (Courtesy of University of Minnesota)

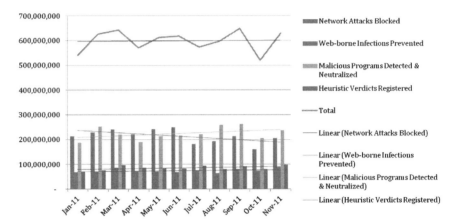

FIGURE 10.4 2011 Threats, according to Kaspersky Lab—Global Research and Analysis Team (GReAT)

retrieved it from the spam folder was sufficient for the attack to be successful through an exploit that used a malicious Flash object[15]

The naiveté of the common end user is the main tool utilized by criminals using e-mail as their vector for attack. Originally, the messages were simple, and often included grammatical and spelling errors, but users were not well educated, and fell for them. As users became more intelligent about potential scams, the criminals had to evolve as well. Phishing tools that are available online now come with complete campaigns that promise a specific "take rate" by users, and provide coaching on how to improve them, such as targeting users from a specific bank, or customers from a specific store.

DOES EDUCATION DECREASE CYBERCRIME IN THE UNITED STATES?

One of the critical pieces to consider in regard to overcoming the impact of cybercrime is the awareness piece. As we mentioned earlier in the book, a recent study conducted in London proved that 45% of female respondents would exchange their passwords for a bar of chocolate, while only 10 percent of male respondents would. However, it still amounted for a large percentage of total respondents.[16] (CRN) With increased education, Americans are less likely to fall prey to phishing and pharming attacks. Many U.S.-based banking institutions, such as Bank of America, have implemented customer education sites to increase awareness of phishing and other types of fraud, with a goal of decreasing the primary risk that is beyond their direct control, since it is at the hands of their customers. In 2007 Carnegie Mellon University actually started an online Anti-Phishing Campaign to train users to spot and prevent victims from falling for phishing scams.[17] Oddly enough, more than 50% of U.S. banking fraud victims, whether phishing or otherwise, decide to change banking institutions after they have been victimized.

Outside the United States, where two-factor authentication schemes are in place for most Internet banking transactions, the phishing schemes become more aggressive. In Chile, we now start to see phishing campaigns focused on LAN Airlines, hoping to trick users with falling for something less threatening than a banking site. It is probably helpful, for the cybercriminals at least, that LAN uses the RUT fiscal ID as the main identification for its frequent fliers. This change in psychological profiling of the phishing target provides insight into how evolved this attack vector is now.

INDUSTRIAL ESPIONAGE AND THE AMERICAN EXPERIENCE

Concurrent with the explosion in use of the Internet by Americans has been a sea change in the reliance by U.S. businesses on computing systems. Since the 1990s, U.S. businesses have come to rely on increasingly complex computer systems for every aspect of getting through the business day. From answering and brokering telephone calls, setting appointments, and even using desktop calculators to the creation of the most innovative new products and services, U.S. businesses simply cannot function without networked computer systems on the desk of every employee not actually on the factory floor. Even on the factory floor, workers are routinely using computer systems to accomplish everything from checking the hourly run rate to calculating complex production methods.[1]

Businesses are no better at securing these essential industrial control and business computers than individuals are at controlling the computers they use for home banking, Facebooking, letter writing and porn watching. To make matters worse, computers have replaced traditional business process management tools to such an extent that, at most businesses with more than 100 people, no one actually knows the true flows of data—the true processes by which business gets done. If you were to ask (and some of the authors of this book regularly do) a group of businesspeople to describe even in the highest-level terms imaginable the derivation and flow through their company of the information they use each day, you would find them unable to do it. Large swaths of most business computer networks are undermanaged, not well monitored, and by any measure, not well understood.

---

[1] Not to mention, of course, the fact that most production lines are today themselves controlled primarily by computer-controlled robots and systems; see, for example, Astrom, A., and B.

Into this environment we now introduce cybercriminals. The target is not necessarily financial data, but rather data that can be monetized. Consider the TJX hack of 2005 through 2007. Hackers had breached the company's network through subverting controls of an under-secured wireless network. Once inside, the hackers observed movements of data that they correctly inferred were part of business processes. When they inserted an extra process, one that delivered large quantities of credit card data to them at regular intervals, no one at the company noticed, because they had not done enough business process and data-flow analysis to be able to detect a business process that was not "normal." This is an example of data that is very easy to monetize being targeted by a criminal group that was, in its organization if not its technology, relatively sophisticated.

Intellectual property is more difficult to monetize than credit card and other account information, but it is presumably of vastly greater value to a much smaller subset of ultimate purchasers. If a company is creating an innovative new product, access to the complete set of records which will ultimately comprise that new intellectual property—such as the collaborative communication among members of the team working on the project, schema, blueprints, CAD drawings, process descriptions, draft patent applications, economic and sales projections, draft and final marketing materials, and other information describing the product, what it does, how it will be sold, and how much it can be sold for—can be an invaluable resource to an unethical manufacturer. We know of and have worked the response to scores of cases in which the theft of this type of intellectual property has sped to market a competitive product, even to the extent that the manufacturer that stole the plans ultimately outsold the product produced by the original innovator.

In 2011, the news that Sinovel, the primary China-based customer of U.S. manufacturer American Semiconductor (AMSC), had stolen from AMSC all of the AMSC intellectual property needed to manufacture and maintain solar and wind products, resulted in an 84% drop in the share price of AMSC,[18] and revealed that the core technology was no longer under the control of the creator. It could be copied at will and sold at a massive discount. That massive discount is possible because firms that steal intellectual property have none of the costs of research and development borne by the creators. As David Etue, who has written about the return on investment of cybercriminals, has posited, "Why spend $40 billion developing [software] when you can steal it for $1 million?"[19]

Unethical manufacturers engaging in industrial espionage to steal IP are typically (though not always) located in countries other than the United States, making prosecution very difficult even when such activity is detected. And often, this activity is simply not detected.

According to Mandiant's M-Trends (the annual report of a company dealing in advanced threats to large companies), the median number of days that attackers were present on a victim network before detection in 2011 was 416.[20] Consider the earlier story about the RSA hack: The fact that a social engineering attack (a plausible e-mail) bore the code that would ultimately be used to steal the key to a security technology used to defend against attackers isn't just richly ironic; it shows that even firms that spend their days thinking about this kind of thing are susceptible and that even when detected relatively early[2] the attack can be devastating.

Compounding the lack of visibility into processes and a widespread lack of understanding of the computer systems and networks on which they take place is the complexity of the tasks undertaken in manufacturing and developing intellectual property. Because of the reliance of the United States on the private sector to develop technologies that are key to national defense, the line between industrial espionage and nation-state espionage—plain old spying—is blurred by these complexities in the context of the U.S. innovation system. It is not just manufacturers of consumer products, energy, or chemicals that are targets of industrial espionage, but firms in the defense industrial base as well.

Consider the 2012 report by the U.S. Senate Armed Services Committee that states that the problem of counterfeit parts in defense tools is "widespread," pointing to more than 1,800 cases of counterfeit parts being discovered in U.S. defense products made between 2009 and 2011.[21] Seventy percent of these counterfeited parts were sourced to China. These parts were used in everything from missile systems to submarines, and blame went to a range of issues including the use of substandard subcontractors. That computer espionage was not responsible for information that advanced these Chinese counterfeiters' product design is inconceivable.

The problem of industrial espionage knows no bounds in terms of industry or industry subset: Any intellectual property on which production is based is targeted. We have responded to cases that include defense industrial base, chemical, energy, and consumer electronics with equal frequency; we know of one case in which the entire design of a consumer product was stolen by a foreign manufacturer, which also stole the completed marketing plans for the product. Months later, the knock-off device was introduced and marketed with the same marketing plans—only the logos and company information had been changed.

---

[2] While RSA claims that it detected and announced the attack within 18 days of its launch, this strains credibility in several key areas. RSA refers to the attack as an "advanced persistent threat"—the industry term widely accepted to mean "an attack which was launched from China"—but the persistence is in doubt; see, for example, Carr, J. (2011) "18 Days From 0day to 8K—An RSA Attack Timeline Analysis." Digital Dao Evolving Hostilities in the Global Cyber Commons Blog, June 2, 2011. Available at http://goo.gl/as5i7. However, even if it was a year, that would still be faster than the average claimed by Mandiant of 416 days from attack to discovery.

There are so many examples of corporate and industrial espionage from the past five years that it's difficult to know where to start describing how often this occurs. And the penalties when caught engaging in these illegal activities can hardly be seen as a deterrent: In 2007, Yonggang Min, a former DuPont employee, was sentenced to 18 months in prison and ordered to pay a $30,000 fine and $14,500 in restitution to DuPont. Min had stolen from DuPont's electronic library (as he prepared to leave DuPont for employment at a foreign competitor, Victrex) more than 22,000 document abstracts and 16,700 full-text PDF files of intellectual property and development communications related to a high-performance film Min had been involved with, and other product information.[22] It was estimated at the time that the total amount of IP stolen represented $400 million in damages to DuPont.[22]

More recently, the *Wall Street Journal* reported that an employee who used a thumb drive to steal key intellectual property from Lubrizol Corp. and deliver it to a Korean competitor comprised a small part of what the FBI estimated was $13 billion in corporate espionage losses from October 2011 to May 2012.[23]

Until U.S. companies begin to address the foundational issues behind corporate espionage—control and understanding of business process, data flow, and intellectual property creation and management—and as long as there are such compelling commercial realities behind the theft of industrial secrets, these problems will grow in severity and frequency.

WHEN THE CRIME IS NOT MOTIVATED BY ECONOMICS

There are times, as with the LLNL example earlier in this chapter, where the main motivator for a crime is not financial. At least, the hacker is not looking to monetize the crime directly. The hacker may be under someone's payroll, but he or she is not the mastermind. Corporate espionage, sabotage, and acquisition of key military information are great motivators for cybercrime as well. Other chapters in this book have identified the histories of cybercrime in China and Russia. However, these stories, as you would expect, often intersect. Attacks such as Titan Rain, Aurora, and other, more recent, attacks show that these military tactics can be applied to the cybersphere as well.

Starting in 1999, Titan Rain would prove to be one of the first discovered uses of the ubiquitous connectivity in the United States against itself. Generally blamed on the Chinese, Titan Rain attacked the military and economic networks with a sophisticated advanced planning never previously seen. Many of these attacks were so successful, due to their unexpected nature, that only when their results were revealed were the attacks known. It would be similar to finding that your lucky coin had been stolen from a bank vault when you received it as change a year later.

SUMMARY

The birth and explosive growth of the online experience in the United States created a perfect opportunity for the creation and growth of malware and other fraudulent activity. A large uneducated digital population, and intellectual capital looking for a purpose created a dangerous combination. As such, America truly became the land of opportunity for those willing to risk creating malware, and to release it in the wild.

References

[1] www.isc.org.

[2] www.isc.org/solutions/survey/history.

[3] www.ncsa.illinois.edu/AboutUs/.

[4] www.llnl.gov/str/pdfs/01_98.pdf.

[5] www.amazon.com/Cuckoos-Egg-Clifford-Stoll/dp/0671726889.

[6] www.ex-mozilla.org/demodoc/demo.html.

[7] http://home.mcom.com/.

[8] http://public.web.cern.ch/public/en/about/webstory-en.html.

[9] http://www.internetworldstats.com/am/us.htm.

[10] Kaspersky, Eugene. Personal Interview. 30 May 2012.

[11] (Courtesy of www.google.com/postini/threat_network.html).

[12] http://www.spamhaus.org/rokso/.

[13] http://www.securelist.com/en/threats/spam?chapter=95.

[14] www.dtc.umn.edu/mints/home.php.

[15] http://www.theregister.co.uk/2011/08/26/rsa_attack_email_found/.

[16] http://www.crn.com/news/security/207400319/survey-women-four-times-more-likely-to-give-away-passwords-for-chocolate.htm.

[17] http://cups.cs.cmu.edu/antiphishing_phil/.

[18] Riley, M., and A. Vance. (2012) "Inside the Chinese Boom in Corporate Espionage." Bloomberg BusinessWeek, March 15, 2012. Available at http://goo.gl/ibHAb.

[19] Etue, D., and J. Corman. (2012) "Adversary ROI: Why Spend $40B Developing It, When You Can Steal It for $1M?" RSA Conference Blog. Available at http://goo.gl/a1UlS.

[20] Mandiant. (2012) "M-Trends, An Evolving Threat." Available at http://marketing.mandiant.com/mtrends2012-sm.

[21] United States Senate. (2012) "Inquiry into Counterfeit Electronic Parts in the Department of Defense Supply Chain." Committee on Armed Services United States Senate, U.S. Government Printing Office, May 2012. Available at http://goo.gl/Gj67S.

[22] Vijayan, J. (2007) "Former DuPont worker gets 18-month sentence for insider data thefts." ComputerWorld, November 7, 2007. Available at http://goo.gl/ZG9us.

[23] Perez, E. (2012) "FBI's New Campaign Targets Corporate Espionage." Wall Street Journal, May 11, 2012. Available at http://goo.gl/rqHD2.

Global Law Enforcement

INFORMATION IN THIS CHAPTER:

- Cybercrime Today
- U.S. Federal Law Enforcement
- Nonfederal Law Enforcement

INTRODUCTION

It's summer 2012, and a quick look at the Cypher Law Group's U.S. Cybersecurity Legislation Tracker[1] shows more than 40 separate federal legislative activities around the creation of cyber security laws in the United States. That's more than 40 independently derived pieces of legislation, each wending its way through the House or Senate, at various stages of life. From House subcommittee[1] to House committee[2]; to Senate committee[3]; to those that languished in committee before being reintroduced[4], many of these acts do indeed contain useful passages and suggestions.

And all of them have aspects of, and a basis in, existing federal and state laws which are in fact quite strong and broad, have either never been or are rarely enforced, and are regularly the topic of declined prosecutions or declined investigations by agency supervisors who understand that those kinds of cases go nowhere unless the issue made the papers, a celebrity was hacked, or the FBI was irritated.

This situation has ample precedent in law enforcement; allow us to walk you through just one analog.

---

[1] e.g., Cybersecurity: The Pivotal Role of Communications Networks" in the House Energy and Commerce Committee, Subcommittee on Communications and Technology.
[2] e.g., The Promoting and Enhancing of Cybersecurity and Information Sharing Effectiveness Act of 2011, a.k.a. "The PrECISE Act."
[3] e.g., the Cyber Intelligence Sharing and Protection Act of 2011, or "CISPA."
[4] Like John McCain's SecureIT Act, which was killed, then reintroduced under Senate Rule 14, permitting the bill to go directly to the Senate floor.

133

It's 1988. Bobby McFerrin's "Don't Worry, Be Happy" and songs by UB40, Billy Ocean, George Michael, and The Pet Shop Boys dominate the Billboard Top 100. A police officer with a master's degree in forensic science who we'll call "Melanie" has just joined a large East Coast police force. Her specialty, blood and body fluids, is not winning her any friends in the crime lab.

> "The career path was that you went into the lab and got assigned to narcotics, where they thought it was easy science and you got the most testimonial experience in the shortest time," Melanie says. "You inject the machine with a solution and it spits out the answer. It's easy to train someone on. And you testify all the time."

If an officer survived that cycle for a while, he or she would move to another area. So, if the officer wanted to specialize in, say, firearms, the officer's bosses would say, "Okay, you've testified 50 times in narcotics and you didn't screw up, so we'll train you for firearms."

> "When I got there," Melanie says, "I had a degree in this stuff, and I'd been trained specifically in blood and body fluids. I didn't go through narcotics. I went straight through to serology--and it was like, 'Who the hell are you? You haven't paid your dues.'"

Melanie brought experience in using newer technologies and techniques, and she understood things such as the nascent DNA processing that was still so new to law enforcement. She would conduct studies to show that what she was doing was valid, and because the sample amounts needed to get results were substantially less than with the methods in use at the time, she could get better cases. In fact, she could use several times less sample to get far more specific results (down to a specific person) as opposed to more sample which provided broad groupings of people. She worked to change the department's and the legal system's thinking.

It was a tough slog.

She was repeatedly told that she needed to work the way everyone else did. She realized that, to get anything done, she needed to trail-blaze.

We've told Melanie's story not just because it eerily parallels that of cyber investigators today, but also because it illustrates three main lessons directly applicable to the current environment in cybercriminal law enforcement. In this chapter, we'll take an in-depth look at those lessons.

CYBERCRIME TODAY

The first lesson to learn is this: In many ways, the same basic frustrations Melanie experienced—those caused by mistrust, a fervent cultural desire to avoid change, an unwillingness to admit that current practices are outdated, a lack of training, and a significant lack of financial resources—are felt keenly by today's cyber cops.

This is a fantastic time to be a cybercriminal. The situation today is analogous to the criminal opportunities that abounded in post-World War II Germany or late 1991 Moscow: The field is wide open, and the authorities have not begun to organize, let alone address, the fundamental issues necessary to combat the unprecedented criminal opportunities presented by a revolution in criminal technology.

Today a criminal of merely average intelligence but above average guile has a choice: Stick up a Kwik-E-Mart, net an average of $653,[2] and face a very high probability of getting caught and imprisoned for several years; or buy a semi-custom malware Trojan, run it for several months, reasonably expect to take home $50,000 when it's all said and done, and face almost no possibility of legal consequences.[5]

Tough choice, eh?

Lest you get too depressed, it's important that you recognize that, as with crime scene forensics, agencies, prosecutors, and courts ultimately will come around to see and embrace new methods of fighting cybercrime. In a decade, cyber-crime will be a significantly different risk equation for cybercriminals. We state this confidently because even though (as we shall discuss) tech-savvy cops have been raising these issues for more than a decade, vast, logarithmic growth in cybercrime victims and hauls is making cybercrime impossible to ignore or explain away for much longer.

Jurisdictional Issues

Speaking to the 2012 National Security Summit in London[3] John Lyons of the International Cyber Security Protection Alliance[4] said, "If we accept for a moment that the vast majority of attacks on our government, businesses, and citizens are orchestrated and carried out by groups outside our jurisdiction, then presumably you will also accept that working together internationally on the identification, investigation, and disruption of cyber attacks must be a key part of defending our national security interests."

---

[5] There is, of course, very little empirical research on this. In 2007 Sam Curry and Amrit Williams gave a great presentation that touched on this kind of calculus--that is, the risk versus reward for a cybercriminal--at www.rsa.com/blog/pdfs/economics_cybercrime.pdf. You can also read several posts on this subject at http://policeledintelligence.com.

In the United States, even in cases where jurisdiction is technically not an issue,[6] no sergeant or administrator from a police agency is going to investigate a crime that was conducted in, say, Los Angeles—let alone Romania—against a local resident for a total loss of, say, $400. It's just not good police economics, and the likelihood of success is incredibly low that the officer would even identify the person responsible. It's inconceivable that, even in the unlikely event of an attribution, an attempt would be made to extradite the criminal from one jurisdiction to another, even if the cybercrime was for $40,000 or maybe even $400,000.

The big problem is that it's rarely $400,000. These kinds of crimes are wholesale in nature, and make what the telecom companies like to call "small ticks"—a couple hundred bucks in jurisdiction A, a few hundred or a thousand bucks in jurisdiction D, and so on. This explains the national policing interest (in the case of the United States, federal law enforcement of various flavors, as we will discuss).

In his summary of the problem, Lyons immediately escalates the cybercrime issue to a national level for this very reason, and as we will explain, this is possibly true in all cases except in the United States, where national policing is more complex and strangely decentralized than in many other countries.

Lyons goes on to say that, "working together internationally on the identification, investigation, and disruption of cyber attacks must be a key part of defending our national security interests," and this is another issue. A cybercrime against businesses or individuals in countries with more liberal socialist policies may take on a different priority than one in more conservative and "small-government" policies, and in surprising ways. American credit card companies typically eat fraud losses—that is, they don't request police intervention for many crimes under a certain threshold—for a variety of reasons which may be summed up on a general level as "the cost of doing business"

---

[6] In the United States, even many local police officers have, technically, investigative prerogative over crimes committed in their jurisdiction regardless of where they emanate. In Texas, for example, technically, any ol' police officer is empowered to investigate on the state's behalf a crime against a Texas person or entity from wherever it is launched. Texas Penal Code, § 1.04 says, in part, "[Texas] has jurisdiction over an offense that a person commits … for which he is criminally responsible if … an element of the offense occurs inside this state; the conduct outside this state constitutes an attempt to commit an offense inside this state; the conduct outside this state constitutes a conspiracy to commit an offense inside this state, and an act in furtherance of the conspiracy occurs inside this state; or the conduct inside this state constitutes an attempt, solicitation, or conspiracy to commit, or establishes criminal responsibility for the commission of, *an offense in another jurisdiction that is also an offense under the laws of this state*." [Emphasis added.] But of course, this is impractical, and of course, it hardly ever happens. The response to a cybercrime in one's jurisdiction is often to either ignore it and hope it disappears under a mountain of other work, or refer it to the FBI and forget about it.

and on very specific levels as "My bonus as chief of security here is to ensure that I ensure the rate of growth in fraud is limited to X percent (or basis points) over the past year"—which means it becomes more costly to investigate fraud and crime than to simply accept it and move on.

The motivation for the prioritization of cybercrime, therefore, varies greatly on national and regional levels, making international cooperation of the sort that Lyons calls for very problematic.

Successful Cooperation

There are clearly examples of successful international law enforcement agency cooperation, and this book does not seek to minimize them. Around the world, agencies are working hard on ad hoc and formal task forces every day on these issues. In June 2012 the Department of Justice announced the arrest of 24 people in 13 countries—including 13 people in the United States—for theft and wholesale of 411,000 compromised credit and debit cards in what became known as the CarderProfit case. That arrests took place in the United Kingdom, Germany, Italy, Japan, Bosnia, Bulgaria, and Norway attests to this high level of coordination and cooperation among international law enforcement agencies dedicated to attacking this kind of crime.

But this kind of coordination is dramatically expensive and difficult to carry out for trivial matters. Time zone differences, international coordination of conference calls and travel, and broad understanding of the nuances of the local laws and customs of each jurisdiction is not something to be undertaken lightly—this is the reason you don't get a three-country national police task force going because a kid stole a TV in the United States, sold it in Canada, and fled to Scotland. So, until recently, this international law enforcement cooperation on cybercrime has been, by necessity, tied to large and financially compelling crimes, or those that have embarrassed certain governments or agencies. As a rule of thumb, in 2012, if your cybercrime doesn't net more than half a million dollars, or does not get in *Time* magazine's international edition or the *International Herald Tribune*, or does not embarrass the FBI, the English or Dutch Hi-Tech Crime Unit, or the *Bundespolizei* in Germany, you're probably not going to be investigated. Indeed, research by the Dutch national police academy was said to confirm this in June 2012:

> "Common forms of cyber crime, such as online stalking, phishing and skimming, are rarely investigated by the police, despite the establishment of a special cyber crime training programme, according to research by the police academy.
>
> The average police officer lacks basic knowledge about cyber crime and the issue is not a priority for public prosecution departments, the report

states. In addition, police experts in digital crime are mainly used for ordinary investigations with an internet element."[5]

This is not exactly a scoop. We and other tech-savvy law enforcement officers have been shouting this from the rooftops for several years now. Consider this quote, from Los Angeles Police Department sergeant and investigator Marc D. Goodman:

> "Computer crime has been recognized as an enforcement dilemma for at least two decades, yet the majority of police agencies seem unconcerned with its presence or effects. Although some strides to investigate and prosecute such crimes have been made recently, the challenges facing the police in their struggle to catch up with the hackers, crackers, and crypto-anarchists of the digital world remain formidable. Despite the recent increase of technology-related crime, 72% of police departments and 88% of sheriff's departments do not have units that specialize in the area."--Marc D. Goodman, *Why The Police Don't Care About Computer Crime*

Now consider that Goodman published these insights in the *Harvard Law Review* in summer 1997.[6]

As Goodman wrote in 1997, cybercrime was increasing, but the ubiquity of Internet access and the reliance on Internet services was nothing like it is today. According to the U.S. government,[7] 18.6 percent of U.S. households in 1997 had Internet access—that is to say, they had dial-up. When one wanted to check e-mail or "surf" the "Information Superhighway," one used a modem to dial into a server and connect at speeds of about 28.8- to 56Kbps. Always-on access to the Internet at speeds of more than 150Kbps—the very ground floor of what we can charitably refer to as "broadband"—wasn't even being tracked by the U.S. government until 2000, when 4.4 percent of U.S. households had it. By 2009, broadband access was enjoyed by almost 64 percent of U.S. households.[8] This means that since Goodman wrote his comments, hundreds of millions more computers are now quasi-permanently connected to the Internet.[7] In that same period, access has exploded to off-the-shelf, commercially available attack and criminal exploitation software that allows criminals to take advantage of the availability of these additional targets and the applications running on them.

So crime has increased dramatically and law enforcement efforts to combat this cybercrime have not. While the FBI was running the CarderProfit sting, the

---

[7] As I write this in 2012, my rural residential area still has no commercially available broadband, and I must spend $500 a month on a T1 line, giving me a mere 1.5Mbps connection. My colleagues writing this book all have cable, DSL, and fiber connections offering speeds of up to 50Mbps for less than $100 per month. Sigh.

successful series of arrests in 2012 which we discussed above, criminals around Europe and the rest of the world were running (and they continue to run) operations of their own, targeting the rich recesses of corporate bank accounts.

Using Trojans and malware, the gangs steal credentials, then make transfers and initiate payments from corporate coffers to a variety of destinations. The moneys are then aggregated and moved using a range of techniques including money mules (sometimes unsuspecting dupes) to comingle and launder ill-gotten gains.

Having worked incidents at corporations in which this very thing has happened, we can state that it is a frequent and diabolically difficult-to-spot occurrence that people steal from corporate bank accounts.

Guardian Analytics and McAfee released a paper in June 2012 describing Operation High Roller, a large and well-coordinated cybercriminal operation. In it, the authors estimated that cybercriminals have attempted at least €60 million (US$78 million) in fraudulent transfers from accounts at 60 or more financial institutions (FIs). The report states that the total attempted fraud could be as high as €2 billion.

What do we take from these simultaneously announced and diametrically opposing operations? The FBI and Department of Justice and all internationally participating agencies should be applauded for proving that, when they work together, law enforcement from vastly different backgrounds can accomplish great things. The fact that the operation netted 24 arrests, and will likely net more arrests in time, demonstrates that even a highly aggressive operation such as this one takes a very long time, involves lots of moving parts, and requires tremendous application of resources, expertise, and coordination.

Meanwhile, the High Roller operation shows that large, well-coordinated, well-financed, and relatively sophisticated organizations are out there working tirelessly on new ways to digitally separate people and corporations from their hard-earned real-world money. They are proof that criminals can move faster and be more innovative than law enforcement, and that in the time it takes to sack up 24 guys, operations large enough to steal €2 billion and avoid capture can be imagined, created, organized, and enabled.

It also demonstrates that cybercrime is not nickel-and-dime stuff committed by kids in their parents' basements, but rather a serious source of illicit revenue engaged in by professional, organized criminal gangs and groups.

This leads us to the United States and its law enforcement objectives. Since the United States is certainly the largest single target of cybercrime, what the country does to combat cybercrime is likely to affect and influence the way cybercrime is fought throughout the world.

U.S. FEDERAL LAW ENFORCEMENT

The FBI continues to aggressively push for dominance in cyber enforcement and they're hopelessly outgunned, despite showpiece roll-ups such as CarderProfit, and high-profile arrests of members of criminal cyber gangs such as Anonymous. The FBI has cybercrime-fighter squads in each of its 56 field offices (each cyber squad can include agents, analysts, linguists, and support staff; obviously some [such as San Francisco] are bigger and better than others), and it says that about a thousand cyber agents do forensics and run operations. According to the Department of Justice Office of Inspector General (OIG), a large number of these are not actually qualified for the tasks they're charged with carrying out.[9] [Thirty-six percent] of agents interviewed by the OIG stated they lacked the networking and counterintelligence expertise to investigate national security computer intrusion cases, and five told investigators they "did not think they were able or qualified" to investigate these kinds of cases.[10]

Regardless of the qualifications (we personally know more than a dozen excellent and totally qualified FBI cyber investigator special agents), the agents and squads are vastly overworked, thoroughly under-resourced, and forced through the realities of these conditions, plus those of politics, caseload, and prosecutorial discretion, to be highly selective about those cases they can accept for investigation. This is not a question of competence, but rather of training—and not training on the job, but a rich tradition of training starting in the academies and continuing with annual re-currency training of the sort all law enforcement officers are required to engage in for issues ranging from racial profiling awareness to traffic enforcement techniques to active-shooter training.

It also doesn't help that the FBI is culturally aggressive in maintaining dominance and terrible at information sharing. What you've seen in the movies is accurate: Cops bristle at the thought of "cooperating" with the FBI which, to cops, means giving the FBI the information the FBI wants and getting little or nothing in return (or, worse, working a complex investigation only to have the FBI usurp it).[8] This means the FBI typically must engage alone or with new allies, because frankly, they burn bridges regularly.

---

[8] See, for example, Geller and Morris, who wrote in 1992 that this Hollywood cliché is accurate, and that local police have long complained of being patronized, alienated, upstaged, and ignored by FBI special agents (Geller, W., and N. Morris (1992) "Relations between Federal and Local Police." *Crime and Justice* 15, Modern Policing (1992), pp. 231–348, The University of Chicago Press).

The Need for Metrics

However, we think the single biggest issue faced by U.S. law enforcement is that it does not have language, or any actual metrics, to define or describe cybercrime. In "Cybercrime: Conceptual Issues for Congress and U.S. Law Enforcement," a 2012 report by the Congressional Research Service,[11] Kristin M. Finklea and Catherine A. Theohary make this case most compellingly:

> "The U.S. government does not appear to have an official definition of cybercrime that distinguishes it from crimes committed in what is considered the real world. Similarly, there is not a definition of cybercrime that distinguishes it from other forms of cyber threats, and the term is often used interchangeably with other Internet- or technology-linked malicious acts. Federal law enforcement agencies often define cybercrime based on their jurisdiction and the crimes they are charged with investigating. And, just as there is no overarching definition for cybercrime, there is no single agency that has been designated as the lead investigative agency for combating cybercrime."

The Connection between a Lack of Metrics and Failure

Without metrics or a definition, there is no money for training (more on training shortly), and without training there's no competency. Without metrics, there can be no determination as to which federal law enforcement agency should in fact be tasked with being the lead in fighting cybercrime. Should it be the FBI? Probably, but not necessarily—there are great cases to be made for the job to go to many other of America's scores of armed U.S. federal law enforcement agencies.[12]

For example, the United States Secret Service is already doing fantastic work; it patently understands the value of proactive defensive measures as opposed to investigative skills after the fact. The U.S. Marshals Service has some serious geek-fu, is highly experienced at finding people, tracking and disposing of ill-gotten gains, and providing protective services. Even the U.S. Postal Inspection Service—arguably America's oldest law enforcement agency[13]—has a totally legitimate claim here: highly trained investigators empowered to investigate crimes that "fraudulently use the U.S. mail, the postal system, or postal employees." Many cybercrimes end up using the postal system (or FedEx or UPS) to transport ill-gotten goods and money. Is it a stretch? Sure. But so is taking 200 special agents and claiming to be the "predominant cyber crime enforcement agency."[14]

The real issue: Without metrics, there's no basis for funding, no basis for this discussion, and the status quo continues. What's lost in this discussion are the 18,000 local, county, state, and tribal police agencies which so far have been largely ignored in America's "fight" against cybercriminals.

NONFEDERAL LAW ENFORCEMENT

The biggest challenges to nonfederal prosecution of cybercrimes are a lack of training, a lack of clear guidelines from prosecutors as to how to investigate cybercrime, and an unwillingness to squander what scant resources are available on crimes which are considered "unimportant"—even, it turns out, when the victims are the police themselves.

Statistics: Law Enforcement in the United States

To discuss nonfederal law enforcement in the United States, we have to give a short background and context. According to the Bureau of Justice Statistics (BJS), by combining local, county, state, and tribal law enforcement agencies in the United States, you get a number close to 18,000 (in 2008 there were 12,501 local police departments, 3,063 sheriffs' offices, 50 primary state law enforcement agencies, 1,733 special jurisdiction agencies, and 638 other agencies, primarily county constable offices in Texas). [15] But this can be highly misleading. About half of all agencies employ fewer than 10 full-time officers and the largest 7 percent of agencies employed 64 percent of all sworn personnel.[16]

By "local" law enforcement, we generally mean city, town, and village, or municipal law enforcement agencies—the local fuzz. There were 12,575 local police departments operating in the United States during 2007, which employed approximately 463,000 full-time sworn personnel.[17]

When we describe county law enforcement we generally speak of sheriffs, though not all states use sheriffs the same way. Typically the sheriff is an elected position and the office is custodian of prisoners taken within the county, for some law enforcement in areas outside municipalities, and in some cases the sheriff provides civil service and enforcement functions as well. In some counties, the sheriff performs all of the above duties, in others some of them. The BJS found that in 2008, there were 3,063 sheriffs' offices with the equivalent of at least one full-time officer, and as a total employed 183,000 full-time sworn officers.[18]

All U.S. states maintain a state police agency,[19] which provides law enforcement, traffic enforcement, patrol, investigation, and forensics, and in many cases, cybercrime or cyber forensics investigative services. In 2004, state police agencies employed 58,190 full-time sworn officers.

In 2008, Native American tribes operated 178 law enforcement agencies that employed the equivalent of at least one full-time sworn officer with general arrest powers; about 3,000 sworn full-time officers work for tribal agencies.[20]

There are a few hundred other county law enforcement agencies, mainly comprising Texas constables. In Texas, the office of the county constable is an elected position, and county constable deputies provide a hodgepodge of civil enforcement, traffic enforcement, and, often, patrol and other law enforcement functions that vary from county to county.

To get a sense of how this plays out locally, it's telling to look at the largest agencies. In 2008, the largest local agency was the New York City Police Department (NYPD) with about 32,000 sworn officers. Next largest was Chicago, with 13,354, followed by Los Angeles (9,727), Philadelphia (6,624), and Houston (5,053).[15] This precipitous drop-off shows that, after the top 20, agencies measure officers in hundreds, not thousands. About 85 percent of all U.S. agencies employ fewer than 25 officers. This means human and technical resources are stretched very thin in all but the very largest agencies. Since cybercrime is not considered to be a priority in any agency we have seen, the availability of personnel to work on and train in cybercrime is, as you can see, a difficult-to-sell prospect to an administrator looking to maximize efficiency and most effectively leverage the few officers he has.

There are exceptions to these statements: Many of the country's largest agencies—and some smaller ones—have dedicated substantial resources to the cybercrime issue. The NYPD, Nassau County (NY), Los Angeles, Miami, Dallas, Chicago, and several other agencies have made priorities of understanding cybercrime. But the impetus, generally, has been for counterterrorism, surveillance, and intelligence and digital forensics and not for garden-variety cybercrimes such as account takeover, Trojans, card skimming, and other digital scams that separate people and businesses from their money. Why? Because everyone understands how to get money for counterterrorism and surveillance and intelligence. To recognize this is to recognize the importance of metrics in a different way.

Additionally, the largest agencies staff their cyber operations not necessarily with cyber geeks, but rather with experienced detectives assisted by cops with some cyber-fu and civilian analysts and other non-sworn staff. This is not wrong: We fully believe that cybercrime is, at the end of the day, merely crime, and that much of the investigation and prosecution comes down to good police work, not good cyber work. Tying a real-world person to a digital crime is a technical process; gathering enough evidence to prosecute, surveilling and observing the suspect, interviewing, and eliciting a confession are the key parts of any cybercrime prosecution, and none of those require the ability to do anything more technical than flipping open a writing tablet and taking manual notes. It's, you know, *police work*.

Training Issues

In a nutshell, law enforcement has no farm team. There is minimal, and often terrible, cybercrime training available in police academies. Cyber is not considered a "real" crime by many cops, who look at it with an enormously dated concept of hacker kids living in their parents' basement—a point of view not helped by the hacker collectives such as LulzSec and Anonymous destructively hacking for glory and for publicity and, on arrest, turning out to be almost the prototypical hacker from the 1980s; *Revenge of the Nerds*, not *The Matrix*.

The "no farm team" comment—made by the administrator of a very progressive agency with an outstanding commitment to cyber—refers to the fact that, traditionally, cops were not high-tech people, and the skills required to be an officer don't typically lend themselves to attracting geeks. Even as this author attended a highly rated police academy in 2010, cyber was a four-hour requirement that we were told by the instructor we'd have to "get through." There is no reserve, no extra capacity in waiting, and no culture of training new recruits in the basics of cyber investigation. Until that changes, law enforcement is playing catch-up.

It's not just cops who need training. District attorneys and prosecutors are woefully unprepared for cyber investigations, because of dated and inaccurate "truisms." For example, most criminal Internet traffic uses devices called anonymizing proxies to obscure the actual IP address. Anonymous proxies, it turns out, present investigators with a major problem, because they present prosecutors with a major problem, because they present a major problem to judges.

It's a problem for judges and prosecutors and ultimately cops because everyone in the criminal justice chain has heard and believes as operational truth that an IP address is like an Internet phone number, telling you exactly where the criminal is—and you can do nothing without it.

The old "IP-address-is-a-phone-number" saw probably began when an exasperated cop was trying, through clenched teeth, to explain something to his super-annuated boss. It has become an indelible myth.

Many prosecutors therefore expect to see the investigator hand up an IP address showing that the computer is in a given location, because it's one of the few places in U.S. cyber law where there's something like precedent: "Of *course* I need the IP address to show the computer at that house," they say, "how else can I *possibly* prove that that computer was used in this crime?"

A wonderful clip from the execrable television show *CSI* epitomizes this.[21] When told that a killer was "online," a "technician" says (in all seriousness), "I'll create a GUI interface using Visual Basic to track the killer's IP address." Sadly, too many in the criminal justice system believe this is real crime-busting stuff.

At the same time, over the past decade or so, cybercriminals have recognized that leaving your real IP address on a server was the single most incriminating thing you could do. In that same time frame, the cost and complexity of obscuring your true IP address has been reduced to, literally, "free" and "one click."

A reminder: Murderers rarely leave white roses, whose thorns bear a DNA sample from where they drew a perfect droplet of blood, and murderers hardly ever leave a calling card with their home number.

It is therefore thoroughly unreasonable to expect that any serious cybercrime will come with a map to the house of the perpetrator, marked with an X where his bedroom computer is.

No, investigators of cybercrimes must often (not always) find other ways of getting attribution, and prosecutors need training. The training needed is both technical and in the area of continuing legal education, because the law changes literally each year, as precedent-setting cases are decided. Prosecutors also need guidance on the evolving face of state law, and federal legislative initiatives, to enable them to "hit the ground running" when new laws are passed that add tools to the prosecutorial toolchest with respect to cyber. Prosecutors must be trained sufficiently to provide their law enforcement agencies and officers with clear, consistent guidelines of what they consider to be the low bar for prosecution: Get me "this," and I can do something. Cops know what a prosecutor needs to prosecute a simple assault; they should know just as palpably what a prosecutor needs to prosecute a cybercrime.

Technology Gap

In addition to the training gap, cops have, to be polite, challenges with their technology. According to the BJS, in 2007 about 9 in 10 local police officers were employed by a department that used in-field computers (up from 3 in 10 in 1990). But the computers are old (generally running Windows XP or older operating systems. This is because law enforcement servers, and the software that's typically sold to local law enforcement is typically green-screen stuff updated for Win32), and typical law enforcement technology does not integrate with other LE technology. The main federal National Computer Information Center (NCIC), maintained by the FBI and used to share information about dangerous criminals with U.S. nonfederal law enforcement personnel, while accessible by IP networks, is based on 1970s-era technology, and despite updates maintains a heroic adherence to obscure, Telex-friendly backward-compatible commands, and data field names and references which belie its COBOL roots.

The main advances in cybercrime technology have been investment in computer hard drive and mobile device forensic tools—notably Encase. Why? Because there are metrics about child pornography and financial crimes committed by computers, and cases against those accused of those crimes are best solved with digital forensics software. Once again, a total lack of cybercrime metrics is behind the failure of local law enforcement to gain and maintain new technologies that could be used to fight, locate, and prosecute cybercriminals.

SUMMARY

To fight cybercrime, U.S. police departments need access to innovative technology in areas including intelligence, analysis, traffic analysis, forensics, and malware reversing. All these tools are part of the standard forensics workbench available to information security professionals in private industry, and all these tools cost money. Nonfederal law enforcement, as we have seen, has no money, and no prospects of getting it—because they cannot state definitively how many victims of cybercrime live in their jurisdiction, nor the value of stolen property through cybercrime, nor the cost of non-enforcement of the poorly defined and terribly understood cybercrime legislation (which has not been explained or effectively turned into specific, consistent marching orders by prosecutors). If you were in charge, would you fund any initiative to combat a generic, nonspecific threat of unarticulated value?

Neither would we.

As we write this in the second half of 2012, the challenges described above dominate the landscape of U.S. cybercrime law enforcement. With training that begins in police academies, with articulation of a universal definition of cybercrime and cyber criminality, and with metrics to measure the success or failure of activities to prevent and prosecute cybercrime, this will absolutely change in the coming five years. Without these activities, the status quo will prevail.

References

[1] https://www.cipherlawgroup.com/legislation.

[2] www2.fbi.gov/ucr/cius_04/offenses_reported/violent_crime/robbery.html; "An average of $653 per robbery was taken at convenience stores. An average of $1,682 per robbery was taken in all other types of robberies, cumulatively".

[3] www.national-security-conference.co.uk/.

[4] https://www.icspa.org/about-us/.

[5] www.dutchnews.nl/news/archives/2012/06/dutch_police_ignore_cyber_crim.php.

[6] [6]. Goodman, Marc D. (1997) "Why The Police Don't Care About Computer Crime." Harvard Journal of Law & Technology 10(3), Summer 1997. Available at http://jolt.law.harvard.edu/articles/pdf/v10/10HarvJLTech465.pdf.

[7] www.ntia.doc.gov/report/2004/nation-online-entering-broadband-age.

[8] www.ntia.doc.gov/files/ntia/publications/ntia_internet_use_report_feb2010.pdf.

[9] www.justice.gov/oig/reports/FBI/a1122r.pdf.

[10] Ibid, page v.

[11] www.fas.org/sgp/crs/misc/R42547.pdf.

[12] http://bjs.ojp.usdoj.gov/index.cfm?ty=pbdetail&iid=4372.

[13] Founded in 1837 by U.S. Postmaster Benjamin Franklin; http://about.usps.com/publications/pub162/pub162_010.htm.

[14] I'm just sayin'.

[15] http://bjs.ojp.usdoj.gov/content/pub/ascii/csllea08.txt.

[16] Ibid.

[17] http://bjs.ojp.usdoj.gov/content/pub/ascii/lpd07.txt.

[18] http://bjs.ojp.usdoj.gov/index.cfm?ty=tp&tid=72.

[19] The BJS has stated that there were 49 primary state police organizations (see, http://bjs.ojp.usdoj.gov/content/pub/ascii/csllea04.txt); this had been expanded to 50 by 2008, as we see in http://bjs.ojp.usdoj.gov/content/pub/ascii/csllea08.txt.

[20] http://bjs.ojp.usdoj.gov/index.cfm?ty=tp&tid=75.

[21] www.youtube.com/watch?v=hkDD03yeLnU.

The Road Ahead

CONTENTS

INTRODUCTION

In this final chapter, each of the authors will provide their insights and opinions on the future of security. The current state of flux in the Information Security arena and the apocalyptic level of changes in the recent past in the Cyber-Terrorism and Cyber-Warfare threaters make it a genuine challenge to look ahead with a high degree of certainty. The authors combined have several decades of experience in security and extremely diverse backgrounds. They have been involved with projects and played roles in organizations which cross the entire information security industry, and range in all levels from practitioner to executive. Additionally, each of the authors has traveled to just about every country in the world speaking and consulting to world governments, financial institutions, retail, defense organizations, electric and utilities to name a few. As you can imagine, John Pirc, Will Gragido, Nick Selby and Daniel Molina, all great friends working for different security organizations, provide their unfiltered assessment of what the road ahead holds for Cyber Crime. They also comment on a few issues within the security community today which act as impediments to any organization worldwide achieving the security needed as they take on the challenges of the road ahead.

You'll find that, while there was violent agreement on general themes, in some cases there was friendly and respectful disagreement on specifics.

JOHN PIRC: "KEEPING SECURITY REAL"

The road ahead for any industry is somewhat hard to predict because there are a lot of unknowns. However, if we continue down the current path of securing corporate and critical infrastructures like we do today, then we are in a lot of trouble and being able to paint the picture of the future becomes easy. I'm sure most of you have heard the saying, "trying the same thing over and over again and expecting a different result"? In my humble opinion, this is exactly the conundrum we are facing today. Most security professionals that have been around the past decade or longer have been talking about SQL injection vulnerabilities and guess what, 12 years later we are preaching the same thing. Have we learned anything in the past decade? I would honestly answer that question with a hell NO! In Chapter 7 I started scratching the surface with introducing the concept of tier 1 and tier 2 security technologies.

Tier 1 Security Technologies

- Firewall or Next Generation Firewall.
- Desktop Anti-Virus.
- Secure Web Gateways.
- Messaging Security.
- Intrusion Detection/Prevention Systems.
- Encryption (in transit or at rest).
- Security Information Event Management.

Tier 2 Security Technologies

- Network Forensics.
- Desktop Forensics.
- Data Leakage Protection (Network/Desktop).
- Network behavioral analysis.
- Security Intelligence Feeds.

Unfortunately, the individuals that are writing "security best practices" and "regulatory compliance" are doing so with technologies that solved a problem a decade ago but do not really address the issues of today and tomorrow. I know this sounds very harsh and likely to upset a few people but it's true. In referencing Chapter 7, the average corporation finds out about an advanced attack by a third party 419 days from inception. I'm sure these organizations followed best security practices, which are mostly tier 1 security products. Again, tier 1 security technologies have there place in the infrastructure but if not augmented by tier 2 security technologies in the right place in the infrastructure and you're a high profile company with intellectual property or information worth in the range of $100,000 - $1,000,000,000 USD, you run a very high risk of being contacted by a third party. Additionally, not every country treats cyber security the same. Not to disclose the specific geographic region but I've visited countries that only had anti-virus and access control lists on their CPE router. What was even more disturbing is that some are just now deploying firewalls…not next generation firewalls but a packet filtering firewall that in my opinion are useless for solving the problems of today let alone the threats of tomorrow. Do you think anti-virus and firewall are going to stop the next Stuxnet and flame malware? Absolutely not, in a very recent breach of an Iranian nuclear power facility, a virus found it's way into their infrastructure and started playing the song "Thunder Struck" by the rock group AC/DC. This is the third time this has happened and although a virus that plays a rock and roll song only signals to that organization and country that whomever is trying to access to their infrastructure has no issues getting inside. This really brings us to the realty and my prediction of what the road ahead holds for security.

2013–2023 Threat Landscape

In the next 10 years it's likely that we will see a cyber attack that will cause a catastrophic fault within a critical infrastructure. It's possible that an attack at this level will lead to an environmental disaster or perhaps human casualties. Stuxnet is proof that this can happen today by allegedly state sponsored actors. Furthermore, the Stuxnet code was released in the wild giving everyone access and the opportunity to study the code. In 2009, I gave a lecture in Brussels on Cyber Terrorism and even hosted a live webinar with the SANS Institute on the topic. At the time, I think the topic was a bit advanced but demonstrated just how easy it was to get into a SCADA infrastructure through multiple vectors. If a terrorist organization or extremist group with right malware and vector into a SCADA infrastructure could cause a lot of damage. This would be no different than strapping on a virtual bomb vest to support a radical belief that was above and beyond the context of any religion or political ideology.

The Rise of Mobility with Wi-Fi/3G/4G

The "bring your own device" movement (BYOD) that is exploding in corporate environments is starting to expand the corporate threat landscape from the desktop to your pocket. In a recent conversation with one of the largest High Tech companies in the world, the Chief Information Security Officer admitted that BYOD is one of their greatest security challenges in terms of controlling intellectual property. I'm a huge fan of BYOD and mobility as it provides me a lot flexibility and accessibility to my email and data anywhere, anytime all over the world. However, most BYOD's on the corporate network are loosely managed and for the most part lack security. In one of the latest high profile malware called Skywiper and Flame had the capability of propagating and infecting other devices via Bluetooth. Just imagine all the Bluetooth enable mobile devices and the use of near field communication. I think in the next 10 years, we are going to see a shift of attack propagation from the traditional wired network devices to mobile devices. Think about it, if I can get access to a mobile device, I can record boardroom and personal conversations, take pictures/video, read email, SMS and download any corporate data that is stored on the phone. This isn't far fetched as I've seen these capabilities being used today but will be a large part in expanding the corporate security infrastructure from the desktop to the pocket.

Changing the Traditional Mindset in Order to Secure the Future

The Information Technology spend on security in most organizations today are typically in the low single digits. The average security spend I typically encounter is around ~3% of the overall Information Technology budget. This can explain why most organizations do not have the budget dollars to expand their security beyond tier 1 security technologies. Additionally, regulatory compliance and security best practices are also key drivers in what security technologies are purchased. These are not all the factors that drive technology buy decisions, familiarity and certification with a specific vendor brand is likely to follow the individual from company to company. The common joke around the industry is that no one ever was fired for buying Cisco. However, if the specific vendor of choice is not investing in security research and just doing enough to stay competitive is a major investment risk. Also, like a 401 K, you do not want to place your entire investment portfolio into one stock or mutual fund. A diverse portfolio typically yields a higher return. The same is true with your security infrastructure investments. Do not place all your eggs in one basket and if you do your research, you will find that small security start-ups and upstarts typically have complimentary security technology (Tier2) that is more advanced and more likely to catch unknown threats on your network that most tier 1 security technologies would miss. I would highly recommend any organization to invest in tier 2 security technologies or at least bring them in for a demonstration. The lack of visibility into abnormal network behavior in any network only raises your risk significantly and most tier 1 security technologies are going to fall short in providing you this information.

WILL GRAGIDO

There is a stark reality facing the world today. It suggests that as the world becomes more and more 'flat' with the concept of distance removed and forever replaced by measurements of time in milliseconds from one host to another, that our lives, livelihood, freedoms and privacy are all subject to scrutiny and exploitation by threat actors of varied denomination. This is in fact, the reality in which we live and operate on a daily basis in 2012. It is one in which there are no guarantees of anonymity in the slightest especially as the denizens of the Internet continue to surrender their rights to said anonymity and/or privacy in order to participate in social networking media, Internet based commerce and entertainment all in an effort to remain and ensure that they remain 'connected'. This is the reality that we all live in; the authors of this book included.

When considering this reality, we must ask ourselves at which point is too much a 'good thing' too much? Is there a point where being perennially connected to the Internet with our personal data on display for the masses in databases, social media networking ecosystems and Twitter feeds simply not good for us? The authors of this book hold a variety of opinions on this matter however, we are all in singular agreement that the likelihood of seeing those participating within the greater digital community in one way or another will remain strong rather than diminish over time more so than not. As a result, it is the belief of the authors that we individually and collectively have a responsibility to be and remain informed of what occurs on the surface, and below the surface of the interwebs that we search and surf on a daily basis. Why you ask should we be concerned? The facts as presented throughout this book and on a daily basis the world over throughout myriad media channels should be enough to answer any who inquire regarding the 'why'. We live in troubled times. We, as a race, have always lived in troubled times. Our ability to see, to hear, to experience that trouble has only been amplified by our ability to remain connected to the Internet much in the same way that our ability to digest, process and arrive at conclusions much more swiftly has also been amplified. What has also been amplified and increased is our susceptibility to exploitation and compromise in ways that in ages past would have been written off as being next to impossible. We know that this is not the case today in 2012 at the time of this writing.

More so than ever before, we have an obligation to ensure that, as individuals, as users of technology, we take as many measures as we have at our disposal to manage and preserve our individual and collective attack surfaces. This is by no means a trivial task especially in an age where remaining 'connected' is in so many ways the norm and to do otherwise is viewed as odd or questionable. How quickly time and attitudes change. As information security professionals and practitioners, we, the authors of this book, believe it is our responsibility to aid as much as we can in spreading this message and the ideas that support it. The road ahead for this generation and the next remains unclear. Technology and those behind pioneering innovation will continue to march on; carrying with them their dreams, ideals, and ambitions so that future generations can enjoy the fruits of their labor much the way that the generation before this one did in the early days of the commercial Internet. So too, will those who seek to exploit the susceptible and vulnerable for profit and gain in the hopes of promoting their own agendas and ideals while achieving their goals. It is for this reason that we must continue to admonish and integrate sound and critical thinking with respect to information security on a daily basis in the lives of the every man. To not do so sees us, the informed, acting in a manner that is complicit with those who seek to exploit and in many ways destroy that which so many have worked so tirelessly to achieve: a world where information can be exchanged in a timely fashion in a manner which allows the every man to access and digest that information through the medium and form factor of his or her liking.

This is in no way a trivial concept. It is, in fact, a weighty one as the stakes are now (and will likely continue to be), greater than any those could have dreamed of in year's prior. For the cyber criminal, the world will remain a bounty cornucopia with fruits ready to yield profit ushering in unprecedented gain. Until and unless, applications developers begin to develop code securely without being told so and network infrastructure and their host (server, end point and mobile) counterparts set out to secure the infrastructure and devices which enable connectivity to the Internet in the most secure manner possible in the absence of the threat of audit or penalty, the cybercriminal will continue to win. He will see the road before himself and his compatriots as being open; without obstruction and in the course of time, he will continue to do what he does best: prey upon those who are unaware of his existence let alone his intentions.

This generation of information security practitioner and those who follow must act in a manner that is equivalent to second nature; they must live and breathe that which is preached in academia and the conference hall. They must seek to promote knowledge and awareness such with respect to the realities of remaining perpetually connected to the Internet at home or on the job; much in the same way that those who traversed the Silk Road hundreds of years ago had to promote that which was safe and known to be sustainable during their expeditions along with that which was unsafe and should be avoided at all costs. Our world may have become infinitely smaller due to the pervasive nature of our desire to remain connected with one another but our sensibilities and awareness does not have to be sacrificed in order to enjoy the benefits while being aware of the risks posed by this change.

NICK SELBY

As cyber-criminals have enhanced their tactics, techniques, and procedures, and as the cost of tools to seek and discover software vulnerabilities and then exploit them have soared in efficacy and plummeted in price, corporate America has remained stuck in a long-passed paradigm. To this day, some of the world's most successful companies continue to rely on the security products of a bygone era to defend against attack. Today's cyber criminal attackers are quite simply years ahead of our defenders.

The future, then, holds some more sophisticated surprises. As defenders up their game, attackers will continue to seek to outpace these increases in defense.

Intelligence Gathering Criminals

We have already begun to see criminal groups and nation states using intelligence to define and target categories of victims. As we saw in the Aurora attacks, and again in 2011 in targeted attacks against companies in the worldwide chemical sector, criminals are leveraging their understanding of how companies (victims) in single industry groups fall universally or generally short in given defensive areas.

To gain this understanding, criminals are engaging in a wide range of traditional intelligence gathering activities. These include

- Human intelligence (HUMINT), the art of infiltrating an organization or hiring a current employee to act as an agent;
- Open source intelligence (OSINT), which is the act of scanning published materials which are openly available, such as newspapers, magazines, blogs, trade publications, conference papers etc, for telling information, data and intelligence on a given subject; and
- Signals intelligence (SIGINT), which is the interception of electronic communications for the purpose of gleaning information or drawing inferential conclusions.

The level of sophistication is relatively high, but by no means has it maxed out. One truth of sophisticated attackers is that they will not spend $100 if $50 will get the job done: they will use the least expensive and least complex solution to accomplish the mission. So because defenders have been so universally bad at mounting a substantive defense, attackers have not had to be particularly sophisticated or clever to overcome the defenses of today. Expect this calculus to change as defenders get better and attackers must up their game to continue to maintain their advantage.

Intel Driven Attacks

As these attackers improve their advantage, expect that the attacks themselves are driven by better intelligence. Today attacks are mounted and if they fail, different attacks are mounted. As defenders increase their capabilities, attackers will need to conduct more intelligence gathering not just on the targets but on the methods of attack, to reduce the noise they make while attacking.

Intelligence Driven Defense

This is not all bad for the defender. What is good for the goose is, in fact, good for the gander, and intelligence is the fastest way to bring the situation to parity. As many have pointed out, one needn't be faster than the shark, just faster than the next fella, and even in an industry which has been specifically targeted for attack by an intelligence-led attacker, defensive intelligence can mean the difference between leaving an open door and leaving a locked door with a "Go Away" sign placed squarely in the center.

Today a relatively small number of leading edge companies engage in defensive and threat intelligence gathering and analysis capabilities. The most important aspect of any such intelligence program at a defender organization is that the intelligence unit be capable of spotting anomalies and gathering both wholesale and specific intelligence on threats faced by the organization, *and* that the intelligence organization have the organizational clout – the swing – to get their findings acted upon in a timely manner. It's no use saying that an attack will be launched at 10 if no one will act upon it until 10:30. Intelligence organizations at defender organizations must be capable of mining a multitude of sources across all the disciplines mentioned above – HUMINT, SIGING and OSINT and also to combine sources and reach across traditional barriers to information and intelligence sharing.

Big data To defend, organizations must use widely disparate, heterogeneous sources of information which have been aggregated and correlated. This would include specific threat feeds and signatures of 'indicators of compromise", but also open sources of relevant data that is effectively mined and exploited.

Breaking silos In addition to external sources of information, organizations must look internally across traditional barriers or silos. It is no longer sufficient to have threat intelligence look at merely the information technology realm, it must dig deeper and wider, into areas of intelligence that affects business, supply chain, physical security, logistics, marketing and sales.

Full Packet Capture Signature-based technologies such as anti-virus, intrusion detection, and prevention, and even some of the more sophisticated kinds, like Damballa and Fireeye, are of severely limited and rapidly decreasing value. The only way to understand the traffic on the network is to engage in full-packet capture and analysis, DNS analysis and other incontrovertible signs of exploitation and exfiltration.

No vulnerability data, Actual Attack Data Defenders must increasingly move away from vulnerability scans and other providers of "potential risk" and look towards empirical data that demonstrates weakness through observed bad activity. Risk cannot be set merely by potential; it must be examined in the context of actual exploitation and weakness as demonstrated by observed actual bad traffic.

Expansion in the Pilfered IP Market

While defenders are shoring up their defenses, attackers are increasing their ability to identify valuable intellectual property and capture that IP which is of specific value. Today, as we have seen, the attackers' advantage is so significant that attackers are able to engage in wholesale theft of data. These hauls in turn must be processed ad analyzed, which is resource-intensive and very expensive: for each Megabyte of data stolen, human analysts must pore through and seek to assess value and salability of each datum or set of information captured. As the market for stolen IP matures, it will make sense to conduct better intelligence pre-theft to enable more efficient capture and exfiltration and more efficient and effective monetization or sale of the stolen IP.

Bleak Outlook

This has been, admittedly, a bleak outlook. For the balance to shift, I believe that it will be necessary both for criminal markets to be exposed, and for defender organizations to see unequivocal financial damage to their bottom line. Until these things happen, the status quo on the defense side will remain. But lessons learned can change minds: it is certain that no one at American Semiconductor will ever again underestimate the danger of IP theft: it lost control of its source code when it was stolen by a Chinese competitor, and lost more than 80% of the value of its stock price when this became public knowledge. The longer it takes other companies of all types to observe this kind of thing, and learn the lessons of them, the more profitable will be the market for stolen intellectual property from American business.

DANIEL MOLINA

The old Chinese curse of "May you live in interesting times" certainly comes to mind in regards to the economics of cyber crime as we look at the road ahead.

In the past 8 months, we have seen a rapid acceleration of cyber-crime tools being weaponized into tools for cyber-warfare. What was previously work of fiction, or merely academic theories, has materialized before our very eyes, and sadly truth is stranger, and substantially more dangerous, than the fiction we had imagined. In a twist of life imitating art, countries have now publicized their involvement in Cyber-war activities in a way that opens up an entirely new and scary frontier. More importantly, we start to see the potential theft or leakage of these cyber-weapons into malware accessible to the mere cyber-criminal.

On June 1, 2012, the international edition of the New York Times published on the front page an admission by the Obama administration that they were materially involved in the development and deployment of the Stuxnet worm. Subsequently, Israel, Germany, Great Britain and others have come forth to claim equal footing in this new and lethal cyber arms race. Israel has explained that the base work for Stuxnet was created by them, and then shared with the United States government.

In fact, The US Department of Defense (DoD) asked for US\$1.3 billion of funding for fiscal year 2012 for its cyber forces, including US\$500 million to build a new joint operations centre for the US Cyber Command (CYBERCOM), which achieved full operational capabilities last November. This is no longer 1983's movie "War Games" with the hacked modem asking "Would you like to play Global Thermonuclear War"? We are now dealing with a serious and tangible global cyber threat that must be addressed by every corporation and every government that is connected.

One of the most concerning aspects of this new reality is that, once we get into multi-million dollar development cycles, with such limited targets, the entire econometric analysis of cybercrime changes. The military industrial complex will warp reality in cyber-weapon development, squeezing out many script kiddies and malware developers in business today. To go up against them, whether in trying to match their offense, or in defense is literally a David vs. Goliath cyber-battle. As well, this will divert budgets into new levels of defense and intelligence gathering. The acceleration of discoveries of zero-day vulnerabilities now that it becomes professionalized as part of cyber-offense will bring us back to the era when the countries with the best utilized military cyber budgets will dwarf those that are content to let others carry the burden of R&D. As such, we will start to see cyber-offense and cyber-defense weapons will again become BIS restricted for trade between countries.[1]

Those that survive in this new cyber warfare reality will be forced to develop infallible, bullet-proof operating systems that cannot be easily compromised due to poor forethought and architecture.

In the same way that simple pistols accelerated into cannons and tanks when taken over by the military industrial complex, we can expect that the new cyber weapons will be substantially more advanced and powerful. As such, they will be substantially more dangerous. The key difference, however, is that, while travelling through the un-policed Internet, the source code for these weaponized cyber threats will now become easily available to anyone with a network connection and a computer. With little knowledge, they could be able to alter the payload of a cyber-weapon, and turn it into a dangerous personal attack vehicle that, in the wrong hands can easily be used for cyber-terrorism, or worse. Once the weapons are controlled by anyone with a basic understanding of programming, the rules of the game forever change. The original weapon design may have been costly and intricate, but the alteration will be as simple as hacking codes to overcome DRMS systems, and that is a future of which we should all be weary.

To put this into econometric perspectives, we should consider that in 1968, as part of his landmark work at the University of Chicago, Gary Becker proposed a rather interesting framework for measuring the economics crime. He postulated that all criminals, whether consciously or unconsciously, make a cost benefit analysis prior to each and every criminal activity. The potential gain is weighed against the potential loss, and only if the gain is greater will the criminal act.

Specifically, in terms of cyber crime, this balance has been markedly in the favor of the criminal for a long time. The lack of web law as addressed previously, the difficulty of prosecution, and the jurisdictional nightmares all created an environment that was fraught with opportunity, and proposed little to no risk of being caught. As such, criminals turned www into the Wild, Wild West. With little chance for being caught, and much less of being successfully prosecuted, many criminals opted to fling their networks of cyber-crime far and wide across the Internet, paying particular attention to use jump-off points and mules in countries that provided a crime friendly environment. The smarter criminals weighed the benefits of improved infrastructure vs. the risk of prosecution, and set up shop in countries like Great Britain and the United States. The infrastructure is solid and reliable, and the risks of prosecution, until recently, were minimal. As prosecutorial pressures have increased, these operations have started to migrate to the smallest islands on paper, while trying to maintain their operations in these infrastructure hubs.

Interestingly, all publications prior to June 1, 2012, addressing this issue have been virtually nullified by the entrance into the cyber forum by the greatest powers in earnest.

SUMMARY

As you can clearly see, the opinions of the authors are based on their disparate backgrounds and views of the world. One of the key takeaways, however, is that every organization that leverages computer technology to achieve its goals, whether directly connected to the Internet or not, needs to strongly revisit its security posture based on the new realities highlighted in this chapter. USB devices, as exemplified in the Stuxnet attacks, can now be the most dangerous devices for a disconnected entity, but the lack of a firewall can nullify the best intentions of those that are connected. The incredibly differing realities for those that use one common Internet, and the dangers inherent in surfing it are rarely properly considered by the general public.

Just as we have emerging economies, and within them different levels of emergence, we have emerging users of technology, and these are not limited by national boundaries. It is this difference in awareness that the cybercriminals of tomorrow will continue to address, much like those who attacked currency fluctuations in the advent of the economic breakdown of the late 1990s. The exploitation by state-sponsored entities in this arena will accelerate the weaponization and exploitation of vulnerabilities, and forever change the economics of cybercrime.

Those entities that insist on using dated philosophies or dated technologies to protect themselves will see that the new weapons-grade cyber threats will render their approach useless. This does not mean that they should embrace only the latest technologies. As many have pointed out, the economics of cyber security have been erroneously based on a "Symptom/Solution" approach, instead of a root-cause analysis of the situation. As long as organizations deploy new technologies to address symptoms, instead of taking a step back to engage in deep introspection to address the root cause of the problem, the cyber criminals will always be one step ahead.

When you focus on the known symptom instead of the root cause, you waste many hours on a fool's errand. This was highlighted on the recent Colombian soap opera "El Patron del Mal" about the life of one of the most famous drug lords of the 1980s, Pablo Emilio Escobar Gaviria. Escobar first started exporting cocaine to the United States inside worn aircraft tires, which were destined for a landfill in Miami; one of his accomplices would go to the Miami landfill and extract the cocaine from the discarded tires. Awhile later, Escobar changed his tactics and began inserting the cocaine packets in the compartment intended for inflatable life vests on commercial airliners. Escobar's crew had infiltrated the company in charge of cleaning the planes and could easily pick up the packets, replace them with actual life vests, and put the cocaine in trash bags that were then sent to a landfill; another accomplice would then retrieve the cocaine from the landfill and deliver it to Escobar's distribution network. It took the DEA years to pick up Escobar's change in tactics, and they continued monitoring worn tire shipments in vain. It is claimed that Pablo Escobar stated, "Historically and mathematically, the criminal will always be one step ahead. While the authorities can only combat the known vectors of crime, we have moved on to new and innovative vectors they have yet to discover."[2]

In the same way, information security professionals that only use known technologies against known attacks will always miss the most targeted and dangerous attacks. Adding new technologies to address each new symptom is not only a foolhardy methodology, but it is also an impossible solution to maintain, either strategically, financially or through adding resources with each new symptom.

As we embark on this new era in information security, we should remember Mark Twain's alleged sage advice: that "history does not repeat itself, but it rhymes". As we look back to the recent past, analyzing the preceding twin to our current couplet, will we be able to forecast economically and take advantage of the upcoming stanza? Does globalization allow for changes in the commanding heights of the economy? Does information become the newest and most important commanding height in a digital world, defining our economic reality?

How will this impact the new cybercriminal economy through multinational corporations? Will the entrance of state entities into the fray warp the past? Can we prevent mutually assured destruction in this new cybercrime reality?

The questions are left before you, the reader, so that you can make better decisions moving forward.

References

[1] Commerce Department Export Controls for Dual Use Technologies - Foreign Policy Controls, Chapter 10, Encryption. (Section 742.15) http://www.bis.doc.gov/licensing/exportingbasics.htm.

[2] "Pablo Escobar, El Patrón del Mal", Telemundo, WSCTV, August, 2012. Television.

Index

Note: Page numbers followed by "*f*" and "*t*" indicate figures and tables respectively.

Made in the USA
Monee, IL
31 October 2019